BEASTLY QUESTIONS

Also Available From Bloomsbury

Tradition and Transformation in Anglo-Saxon England, by Susan Oosthuizen
Shaky Ground, by Elizabeth Marlowe
The Archaeology of Race, by Debbie Challis
The Anthropology of Hunter-Gatherers, by Vicki Cummings

BEASTLY QUESTIONS

ANIMAL ANSWERS TO ARCHAEOLOGICAL ISSUES

Naomi Sykes

Bloomsbury Academic
An imprint of Bloomsbury Publishing Plc
B L O O M S B U R Y
LONDON • NEW DELHI • NEW YORK • SYDNEY

Bloomsbury Academic
An imprint of Bloomsbury Publishing Plc

50 Bedford Square
London
WC1B 3DP
UK

1385 Broadway
New York
NY 10018
USA

www.bloomsbury.com

BLOOMSBURY and the Diana logo are trademarks of Bloomsbury Publishing Plc

Hardback edition published 2014
This paperback edition published 2015

British Library Cataloguing-in-Publication Data
A catalogue record for this book is available from the British Library.

ISBN: HB: 978-1-47250-675-7
PB: 978-1-47426-069-5
ePDF: 978-1-47251-494-3
ePub: 978-1-47250-624-5

Library of Congress Cataloging-in-Publication Data
Sykes, Naomi Jane.
Beastly questions : animal answers to archaeological issues / Naomi Sykes.
pages cm
Includes bibliographical references and index.
ISBN 978-1-4725-0675-7 (hardback) – ISBN 978-1-4725-0624-5 (ePub) – ISBN 978-1-4725-1494-3 (ePDF) 1. Animal remains (Archaeology) 2. Human-animal relationships–History–To 1500. 3. Social archaeology. 4. Animal remains (Archaeology)–Europe. 5. Human-animal relationships–Europe–History–To 1500. 6. Social archaeology–Europe. 7. Europe–Antiquities. I. Title.
CC79.5.A5S94 2014
930.1'0285–dc23
2014009087

Typeset by Newgen Knowledge Works (P) Ltd., Chennai, India
Printed and bound in Great Britain

To the cats, chickens and children, and to that bloke who tolerates living with so many girls. Adela, Bronte, Fanny, Finch, Florry, Furkin, Gunter, Ione, Little My, Marley, Mimble, Mimi, Perdix, Pidge, Primula, Peddler, Star, Stitch and Waggers.

CONTENTS

LIST OF ILLUSTRATIONS

Figures

Tables

PREFACE

Despite this book's title, I don't actually have any answers. This volume is essentially a collection of my musing, and the interpretations and ideas that I present are just that: my interpretations and my ideas – they are not to be trusted.

The only thing in this book that I genuinely believe is that the answers are out there, preserved in the animal bones and associated material culture of the archaeological record. When contextualized within wider interdisciplinary research, which to my mind is the only way forward, zooarchaeology has the potential to transform our understanding of the past, as well as the present and the future. Are (zoo)archaeologists up to the challenge? That is, perhaps, the biggest question. But I hope the answer is 'Yes!'

ACKNOWLEDGEMENTS

I wrote this book partly because when, in 2010, I started teaching a new module called 'Beastly Questions', there was no single volume to which I could refer my students. Collectively, the class of 2010 and I took the decision to write the book that we wished we already had. For their assignments, the students undertook original research on any animal-related topics that interested them, be it dragons or wool. Since that seminal year, all the students that have taken the module have done likewise. Already some of these students are rising stars in the field of zooarchaeology (you know who you are) and so I have resisted the temptation to include their work in this volume (they will publish it themselves). However, each and every student, be they stars or slackers (again, you know who you are), helped to create the research dynamic that has made my job joyful for the last four years.

These wonderful individuals – students turned colleagues – must all be acknowledged. So, my thanks go to: Laura Amos, Stuart Ashby, Jessica Beaver, Jonny Bell, Charlotte Bell, Ben Billingsley, Laura Bishop, Katie Blyth, Sholto Bonham, Harriet Brown, Seraphina Brown, Jonny Bulcock, James Burrows, Nigel Byram, Hannah Chisholm, Ben Cockle, Alex Connock, Katie Cooper, Will Coren, Charlotte Cowderoy, Dan Crampsie, Kathryn Dally, Ben Davies, James Dearson, Sean Doherty, Adam Douthwaite, Ilaria Falqui, Liz Farebrother, Andrew Foster, Mark Fussey, Martina Gagin, Nick Gill, Josh Gottlieb, Jasmine Gray, Alix Green, Erin Green, Sam Hall, Alex Hamilton, Becky Hankinson, Hillery Harrison, Jimmy Harthill, Nat Hitchins, Poppy Hodkinson, Sara Holm, Jessica Hughes, Amanda Hurry, Chris Jones, Tomas Joseph, Zoe Knapp, Andy la Niece, Jamie Lee, James Longstaff, David Lucy, Elspeth McKellar, Amandeep Mahal, Lizzie Manchester, Harry Mansfield, Joel Markham, Luke Martin, Sasha Mclachlan, Ryan Moore, Robert Motamed, Chris Nacca, Tania Newman, Rebecca Nightingale, Max Ogden, Fran Patrick, Matt Platt, Rebekah Pool, Rachel Potter, Abi Price, Dale Prime, Sophie Pye, Marcus Rizzo, Duncan Robins, Declan Robinson, Conor Ryan, Kushboo Sagar, Thoman Sanville, Nicholas Scott, Catherine Shaw, Oliver Slattery, Ashley Stabenow, Charlie Syson, Sezin Tanner, James Taylor, Emma Teboul, Amy Tompson, Jake Thorton, Sarah Tilley, Marios Tjirkali, Hilary Tricker, Stuart Tyrer, Victoria Walker, Lee Walkington, Tara Wallace, Rachel Walsh, Hannah Ward, John Watson, Westy, Kate Whiston, Robert Wilford, Lucy Williams, Callum Wilson, Dan Wojcik, Luke Wollett, Annabell Zander, Yiheny Zhou.

Beyond the Beastly Questions students, there are other individuals whose generosity of ideas (Mark Pearce, who suggested that I teach the module in the first place) and data (Martyn Allen, Emir Filipovic, Henriette Kroll, Jim Morris, Kris Poole and Rebecca Reynolds) were vital for this volume. Others kindly provided images: Robin Bendrey, Will Bowden, John Fletcher, Chris Graham, Richard Jones,

Joy McCorriston, Paul Morris, Terry O'Connor, Jessica Pearson, Peter Popkin. My thanks go particularly to David Taylor, for drawing Figures 3.3a and 3.3b, and to Holly Miller for constructing Figure 1.7. Holly also kindly read and commented upon sections of the volume, as did Nina Crummy, Phillip Shaw and Terry O'Connor (who courageously reviewed the whole thing): I may not have taken all your advice but I thank you for offering it.

The research in this book was supported by two AHRC grants: AH/I026456/1 and AH/L006979/1.

Chapter 1

ANIMALS AND PEOPLE: MIRRORS AND WINDOWS

Zooarchaeology has begun to bore me. If I find myself reading, or for that matter writing, another animal bone report where the major conclusion is that an assemblage 'contained 64 per cent cattle and less than 1 per cent deer, suggesting that people ate a lot of beef but that hunting was not very important' (choose any of my reports and a similar statement will be there somewhere) I may have to commit suicide, academically speaking. In fact, this book is probably a major step towards achieving my scholastic death wish, as it presents a very personal, and perhaps not widely held, view about the aims and potential of zooarchaeology: my feeling is that we can do better and may need to if the discipline is to remain viable and respected.

For me, zooarchaeology is the study of animals – their remains, representations (artistic, linguistic or literary) and associated material culture – to examine the most fundamental issues concerning past societies: how people behaved and how they thought. Traditionally, zooarchaeologists, including myself, have shied away from tackling these big questions, instead passing the task, along with quantities of economic and environmental data, over to the 'real' archaeologists. I should, perhaps, not tar the whole zooarchaeological community based on my own shortcomings, as there are certainly many wonderful examples where zooarchaeologists have been at the forefront of archaeological research, asking cutting-edge questions about mainstream issues (e.g. Albarella and Serjeantson 2002; Arbuckle 2012; Barrett *et al.* 2004; Bartosiewicz 2003; Benecke 1994; Clavel 2001; MacKinnon 2004; O'Connor 2001; Outram *et al.* 2009; Pluskowski *et al.* 2011; N. Russell 2002; Twiss 2012; Vigne 2011; Zeder 1991, 2005). However, my impression is that these insightful works are in the minority relative to the large amount of zooarchaeological research that is undertaken worldwide. The situation is brought into relief further by the fact that some of the most exciting animal-based studies have been produced by scholars who would not classify themselves as zooarchaeologists (e.g. Conneller 2004; Crummy 2013; Fletcher 2011; Gardiner 1997; Hamerow 2006; Jennbert 2011; Larson 2011; Whittle 2012). For this reason, because faunal-remains specialists often stop short of interpreting the data they work so hard to produce, zooarchaeology is widely considered to be a facile data-generating specialism that provides little information beyond 'what people ate'. As a result, funding for zooarchaeology is cut, particularly within the commercial sector, fatally reducing the specialist's ability to provide any interpretation beyond mere dietary and economic reconstruction.

This problem is perhaps not helped by the university system. In the UK students are, quite rightly and vitally, trained to identify and record bones and teeth, to collate, analyse and present their data (64 per cent cattle, 1 per cent deer . . .), and to decipher any observable patterns (dietary preference for beef. . .) but they are not generally required to address bigger social and cultural questions about their material. And how could we expect them to do so? Time is limited, the literature that might feasibly help is dispersed widely and, by comparison to the many laboratory manuals for bone identification (Bocheński and Tomek 2009; Cannon 1987; Cohen and Serjeantson 1996; Hillson 1992; Pales and Lambert 1971; Schmid 1972, Walker 1985) and text-books on analysis (Chaplin 1971; Cornwall 1956; Davis 1987; O'Connor 2000; Reitz and Wing 2008), there are few volumes dedicated specifically to the interpretation of zooarchaeological material: N. Russell (2012) is a notable exception, her important volume published whilst I was writing this book.

In the absence of accessible texts it has been difficult to teach 'social zooarchaeology', so cohort after cohort of zooarchaeology students graduate without really understanding the true value or great potential of the data at their disposal. I was one such graduate. At university I received exemplary tuition in the theory and methods of animal bone analysis, trained to a standard where I was a competent commercial specialist. I spent several years producing data-heavy/interpretation-lite faunal reports before becoming a university lecturer, a role that enabled me to equip new generations of students with the ability to follow in my tedious footsteps. I would now like this cycle to stop. So, I have written this book. I hope it will encourage reflection on the core methods and ambitions of archaeological animal studies and highlight the potential for zooarchaeologists to make a contribution not only to archaeology but also to disciplines outside our own field.

The discussion presented here is centred primarily on my own research, which focuses on British material dating from the Iron Age to the post-Medieval period, but the approach I advocate is, I believe, applicable to any assemblage from any period or place. My overarching aim is to encourage zooarchaeologists to take control of the evidence at their fingertips, to demonstrate to the archaeological profession that zooarchaeology is a varied and highly skilled discipline that provides vital evidence to answer significant questions about any culture under consideration. Essentially, animals should be at the centre of archaeology's research agendas

To take a stand, however, requires that we have something to stand upon. Fortunately, zooarchaeology has exceptionally strong foundations, built during the decades of Processual Archaeology, when many of the discipline's most important methods were developed and influential texts written (e.g. Binford 1978; 1981; Brain 1981; Grant 1982; Legge and Rowley-Conwy 1988; Payne 1973; Shipman 1981; Speth 1983). These cornerstones of zooarchaeology have stood the test of time but, as with all great works, they now require some renovation. In the following sections I will contemplate some of the fundamental principles and methods of zooarchaeology and consider if it is time to take a new perspective on some of our approaches.

Step One: Considering the Bones

Figure 1.1 shows an assemblage, the kind that might be encountered on any archaeological site. With the use of a reference collection and laboratory manuals it does not take long to identify that all of the specimens in the assemblage are bird bones and, more specifically, that they belong to domestic fowl (*Gallus gallus*).

Closer inspection indicates that most parts of the skeleton are present – head, vertebral column, wings, legs and feet – and that, in all probability, the specimens come from a single individual: in no case are there more elements than would be expected from a one skeleton. Looking at the two tarsometatarsi, cock-spurs are absent, suggesting that the individual is probably female, and we may assume that it was an adult as the bones are all fully developed. Taking the analysis further, measurements of the bones highlight that this particular chicken is very large indeed, most probably the product of a programme of selective breeding for carcass size. The idea that the animal was raised for meat is perhaps borne out by the cut marks visible on several of the bones, most probably made when the carcass was disarticulated for consumption. That the whole skeleton was deposited together, despite the fact that it was butchered for food, could indicate that the individual was afforded special treatment and may reflect the refuse from a religious-ritual feast. Why this might have occurred is less easy to determine but it

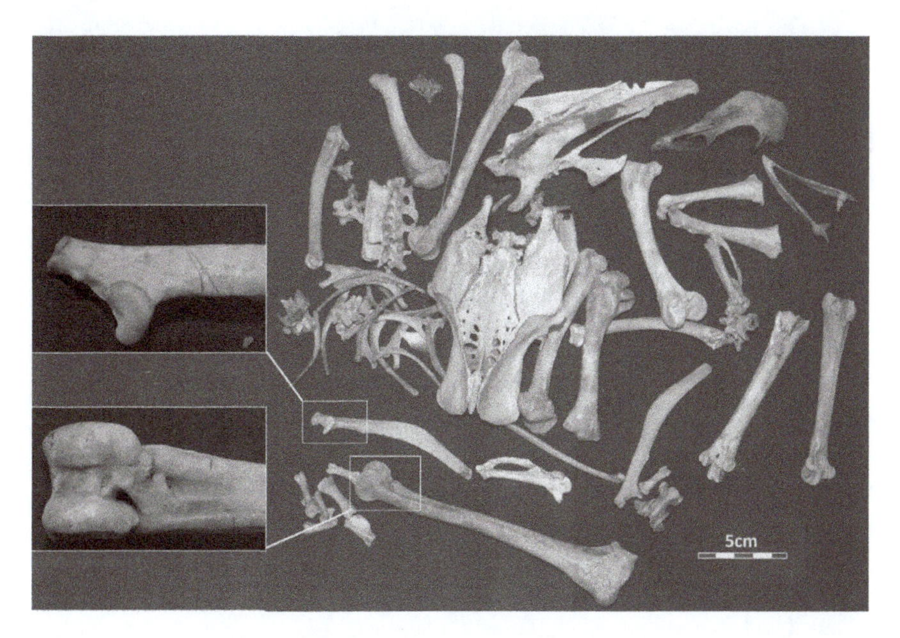

Figure 1.1 An assemblage of very large chicken bones. Several show evidence of butchery (see insets).

Source: Sykes.

could relate to the 'exotic' status of domestic fowl: as with most domestic animals, they are not native to many of the areas of the world where they have been imported (Chapter 4; Sykes 2012).

As far as the above interpretation goes, which is probably as far as is found in most zooarchaeological reports, it is largely accurate, although the chicken is actually a male rather than a female. I know this because this particular 'assemblage' of bones comes from a cockerel, Gunter, who I lived with for almost a year (Figure 1.2).

Gunter hatched on 14 June 2010, matured into a magnificent cockerel by October that year but by February 2011 he had become so noisy and sexually

Figure 1.2 Gunter the Magnificent.

Source: Paul Morris.

aggressive toward the hens that we decided to eat him (to legitimize killing him). As a culinary experience Gunter was memorable for the wrong reasons, reaching unusual levels of toughness: had we wanted him to be less chewy we should have killed him many months earlier. Most of my memories of Gunter are of the time we spent together whilst he was alive, when his beauty, sound and overly virile behaviour was hypnotic, though often disturbing. The relationship that my family and I had with Gunter was very real and very different to the relationship that we had with another cockerel, Nightshade (also recently dispatched). In the case of both Gunter and Nightshade we valued them more in life than in death and I would go so far as to say that the two cockerels, together with the hens – Marley, Mimble, Little My, Mimi, Bronte, Perdix and Pidge – have changed my family's life, lifestyle, diet and our local landscape. For instance, these free-range creatures have not only altered how our garden looks, sounds and smells but they have also transformed the way that we engage with the space on a daily basis. We now spend time looking for eggs, dodging dustbowls and avoiding or collecting the huge amount of shit that they produce. They have even changed the life of our cat, Waggers, who has socialized with them sufficiently to overcome his original desire to eat them, restraining himself even when, as fluffy chicks, they snuck into the house, pushing past him to eat from his food tray.

Spending time with our chickens and cat, and observing their impact on my own behaviour, made me realize that my relationship with these animals was far more complex and interesting than anything I attributed to archaeological assemblages. It made me appreciate the work of Armstrong Oma (2010), Brittain and Overton (2013) and Mlekuž (2007) who have argued that human–animal interactions transform both parties, the two becoming mutually socialized through their exchanges. This has made me reconsider the way I think about and interpret zooarchaeological material. Critics will argue that I am anthropomorphizing the animals that I live with, attributing them with names, characters and thoughts of which they are oblivious. Critics will believe that I am a 'bunny hugger' with little understanding of the kind of relationship that 'real' farming communities, both now and in the past, have with animals.

To some extent these criticisms are fair and I accept that my relationship with Waggers the cat, Gunter and the hens (all of whom will reappear throughout this book) and my experiences of living with them are unlikely to be representative of the relationships and experiences of other people in time and space. The insights I have obtained are largely personal, reflecting my individual worldview. But the fact that my relationship with the chickens is such a close reflection of my wider life concurs with the growing body of evidence from human–animal studies, which highlights that the way that people think about and treat animals is a 'mirror' of, or provides a 'window' into, how they treat and think about each other (e.g. Mullin 1999). I am aware that scholars such as Mullin no longer subscribe to the mirror/windows metaphor, instead seeing human–animal relationships in terms of 'mutual becomings'; however, I like the mirror/windows concept and will use it as a central device throughout this book.

Whatever their theoretical stance, anthropologists, sociologists and cultural geographers all recognize that human–animal interactions are a key source of information for understanding human societies and cultural ideology (e.g. Bulliet 2005; Ingold 2000; Knight 2005; Mullin 1999; Philo and Wilbert 2000; Wilkie 2010; Wolch and Emel 1998). Given that other disciplines appreciate the value of animal studies, archaeologists have been remarkably slow to come to a similar recognition. The situation is made stranger still given that, on many archaeological sites, animal remains are the most common find. In fact it could be argued that the bulk of the archaeological record is composed of debris from human–animal interactions, be it in the form of animal remains, animal-related artefacts (e.g. horse gear, cooking pots, butchery tools, animal figurines) or entire landscapes.

I wonder if our failure to extract decent social information from archaeological animal data is due, in part, to the rationale behind current zooarchaeological methods: whilst the techniques of quantification, ageing and sexing are essential to zooarchaeological work, most were created to ascertain the productive rather than cultural significance of animals. It would take, however, just a small shift in mindset to gain a less familiar but more engaging perspective on past societies.

Step Two: Considering the Methods

It is important to stress from the outset that although this book is not concerned with 'practical' zooarchaeology in the sense of identification and laboratory methods, I agree completely with Reitz and Wing (2008: ix) when they state that 'theoretical interpretations are no better than the methods used to develop supporting data. It is as necessary to be well-grounded in the basics as it is to be guided by good theory'. It is imperative that any data collection is undertaken to the highest possible standards and I would refer the reader to a good reference collection and the works of Reitz and Wing (2008), Davis (1987) and O'Connor (2000) for guidance about how field and laboratory work should be conducted. Due consideration must also be given to issues of taphonomy and all researchers should have Lyman (1994) in their library. Once the data are collected, however, there is scope for re-thinking how we consider results derived from them.

Taxonomy: The Classification and Categorization of Animals

For me, one of the most unnerving aspects of zooarchaeology is taxonomy. The scientific classification system to which we all adhere – namely the Linnaean system – underpins and dictates the direction of all our zooarchaeological analyses, providing neat and internationally recognizable categories such as *Sus scrofa* or *Lama glama* into which we can drop our animal remains. From the perspective of methodological standardization, this is a very good thing. However, given that the Linnaean system is a product of the eighteenth century its applicability to societies that pre-date its invention is questionable. Even in our society, few of us in our day-to-day life actually categorize animals based on the Linnaean system: according to my youngest daughter's favourite book, 'Chick says "cheep!" And I can too! Cheep!' nowhere does it mention *Gallus gallus*. Here we are dealing with 'folk taxonomy', the methods of classification that reflect our bodily perceptions, culture, linguistic traditions and immediate needs; as such they provide excellent information about human belief systems. Folk taxonomies do not always match scientific classifications and there are many studies providing examples where the two do not correspond at all. For instance, Morris (1998: 140) explains that in Malawi, as in many other cultures, there are no terms that are equivalent to 'animal' or 'plant', and both Marciniak (2005) and Wapnish (1995) provide ample anthropological and archaeological examples where animals have been classified not on the basis of any genetic similarity but according to their age, size, colour, whether or not they will be used for sacrifice, the list goes on. As if this was not distressing enough, Marciniak (2005: 57) quotes Foucault's (1970: xv) often-cited extract from a 'certain Chinese encyclopaedia' which discusses animal classification:

> Animals are divided into (a) belonging to the Emperor, (b) embalmed, (c) tame, (d) sucking pigs, (e) sirens, (f) fabulous, (g) stray dogs, (h) included in the present classification, (i) frenzied, (j) innumerable, (k) drawn with a very fine camelhair brush, (l) *et cetera*, (m) having just broken the water pitcher, (n) that from a long way off look like flies.

Like Foucault, I find this extract terrifying as it threatens to nullify every piece of zooarchaeological work that I have ever undertaken. Based on the encyclopaedia, it appears that we zooarchaeologists have fallen at the first hurdle: the Linnaean classification upon which we rely would seem to be the very thing prohibiting us from gaining a clear understanding of the societies we study. By using the Linnaean system we have, from the start of our analysis, imposed our own classification system that will obscure the systems employed by past societies. Marciniak (2005: 54) argues that, if we are to appreciate the social and ideological meaning of animals in past societies, and by extraction the societies themselves, it is first necessary to examine their classification systems. This seems an impossible task, particularly for the Prehistoric period, which may be the reason why few but the most intrepid zooarchaeologists (e.g. Marciniak 2005; O'Connor 2013a; Serjeantson 2000; Wapnish 1995) attempt to deal with the issue: Reitz and Wing (2008: 32–3) devote just one page to folk taxonomies and the issue is overlooked entirely by N. Russell (2012). As much as I would also like to avoid the topic – as the literature on folk taxonomy is expansive and difficult to navigate – I believe it is important to think about our methods of classification. If nothing else, reflecting upon the Linnaean system serves as a useful reminder that, whatever our interpretations, they will always be grounded in our own belief systems, something that we cannot get away from and need to be honest about.

But perhaps all is not as dark as it seems and, to demonstrate this, I will start by clearing up the troublesome matter of that 'certain Chinese encyclopaedia'. Foucault never actually saw the text; he was citing a quotation he read in Borges' 1942 essay *The Analytical Language of John Wilkins* (reprint 1999: 231). In turn, Borges was quoting the work of Dr Franz Kuhn, a German sinologist. Although Kuhn was indeed a translator of Chinese literature and is said to have discovered the encyclopaedia known as *The Celestial Emporium of Benevolent Knowledge*, current evidence suggests he had never seen it either, because it did not exist – he made it up (Longxi 1998: 22).

The encyclopaedia's lack of existence is strangely comforting but it does not remove the fact that different societies do have different ways of classifying animals, some of which are far closer to *The Celestial Emporium of Benevolent Knowledge* than they are to the Linnaean system (again see Marciniak 2005; Morris 1998; Wapnish 1995). However, Morris's (1998: 120–67) study on folk taxonomy concludes that, for all the variations, humans are remarkably consistent in their overarching methods of classification: across cultures people tend to develop animal groups that correspond to 'mammal', 'bird', 'fish', 'worm/snake' and 'others' (e.g. insects, molluscs and crustacea), with different animals being identified within these groups (Morris 1998: 167).

This consistency should offer reassurance to zooarchaeologists, even those working with Prehistoric material. The news is even better for those studying the historic period, especially in Europe. For although we use the Linnaean system, human desire for scientific classification has a considerable time depth and influential antecedents can be traced back to the work of early Greek philosophers such as Plato and Aristotle (Jones 2013: 12). For over two millennia, therefore, animal classification in Europe may have borne some semblance to the systems used by zooarchaeologists today. True, there are plenty of instances where the application of Linnaean classifications would completely obscure folk taxonomies of the past, such as the medieval perception of beavers, barnacle geese and foetal animals as 'fish' and whales as 'monsters' (Szabo 2005) or the widespread incorporation of mythical creatures, such as unicorns, into animal systems (e.g. Pluskowski 2005; Wapnish 1995). But overall, these cases would appear to be the exception and when we analyse the archaeological remains of, say, a red deer (*Cervus elaphus*) and a wild boar (*Sus scrofa*), we can probably assume that the animals from which they derive were understood by the people responsible for incorporating their remains into the archaeological record to be different kinds of animals.

Whilst our ability to have some confidence in our methods of classification is certainly positive, it brings us no closer to the social and ideological meanings of animals sought, quite correctly, by Marciniak (2005: 54). This still remains a difficult task, because although an animal may be classified into a certain animal group (e.g. red deer), the meaning attached to the particular animal will change according to context, even within a single society on a single day. I believe, as Morris (1998: 139) has argued, that comprehension of meaning can only be achieved if we contextualize our data with respect to cultural and social practices. To identify such practices, however, requires us to re-think some of our other core techniques.

Quantification

Desire to determine how well represented different animals are in assemblages has always been at the heart of zooarchaeology. There are many methods of calculation – e.g. NISP, MNI, MNE, MAU, each acronym representing a variety of different sub-methods – and a huge literature surrounds the subject of their calculation and validity (see Lyman 2008). Although quantification methods have come under intense scrutiny, less critical thought has been applied to the meaning of the results that they generate: what do they actually represent in social terms? Reading zooarchaeological literature the obvious conclusion to draw is that we perceive that 'bigger is better' – the more a taxon is represented, the more important it was to that society. In report after report, volume after volume, we devote most of our efforts to ascertaining the relative frequency and role of the best represented taxa. Where species are poorly represented they are frequently labelled 'unimportant'; where they are absent, they slip from our consciousness altogether, even when we are confident that they were available to the human groups we are studying.

It is unfair to be too critical of this situation as it is the product of the zooarchaeologists' necessary obsession with sample size: 'small samples', I warn my students, 'are not statistically significant and therefore it could be dangerous to make too much of them'. But is our concentration on the most abundant, statistically significant taxa always telling us what we want to know? High levels of archaeological representation need not equate to an animal's social importance; indeed, the correlation may be negative: those animals engaged with or consumed on a daily basis may have carried less social significance than animals encountered infrequently (Figure 1.3). This is exemplified well by the fact that, in many past

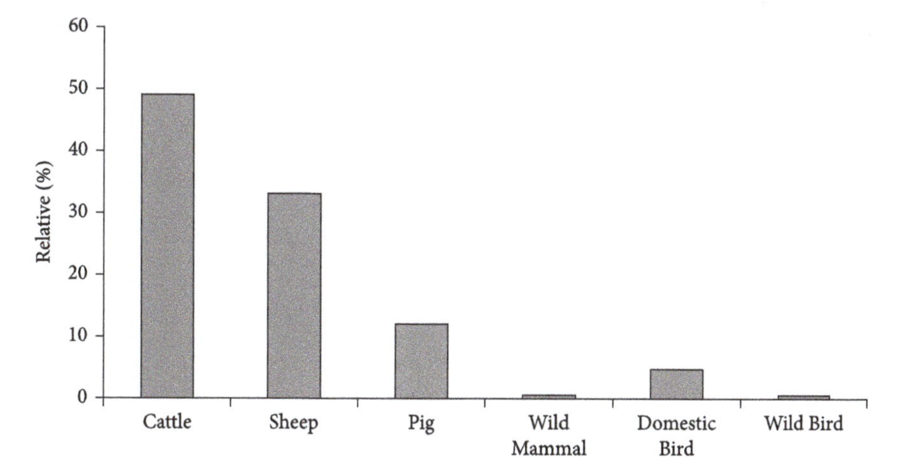

Figure 1.3 Quantify this. Which is the most important taxon? Possible answers: Cattle – their bones are the most numerous reflecting the taxon's centrality to the culture; one of the pigs (a boar called 'Frank'); the bears that are not present in the assemblage but are found everywhere in the art, stories and place-names of the culture.

Source: Sykes.

cultures, hunting scenes figure large in art and literature but the remains of wild animals are scarce on archaeological sites. This apparent disparity between the different sources of information, which is often seen as very perplexing (e.g. for a detailed discussion of the evidence for Roman Britain, see Cool 2006: 111–18), disappears completely if the importance that zooarchaeologists attached to frequency is inverted. I am not suggesting that the formula 'least represented = most significant' is valid either, only that it is important to recognize that the act of quantification is not the end of the journey in terms of zooarchaeological interpretation – it is the beginning. We may be able to quantify the number of bones in an assemblage and extrapolate from this the ratios of livestock or the dietary contribution made by each, but social quantification is an entirely different matter, requiring contextualization and thought.

Skeletal Representation

A similar point may be made with regard to the interpretation of skeletal representation data: the social significance attached to animal body parts in the past may bear no relation to modern perspectives. It is well recognized by zooarchaeologists that different parts of an animal carcass will be deemed variously as 'good' or 'poor' depending on cultural attitudes (Crabtree 1991a: 172–5; Ashby 2002: 44). For instance, whilst in modern England the shoulder and leg of lamb are prized cuts, deFrance (2009: 123) refers to the Druze villagers of the southern Levant who prize the fatty sheep crania over other more 'meaty' parts of the body. Table 1.1 shows how animal carcasses are divided and redistributed by the modern

Table 1.1 Mechanisms of meat redistribution amongst the Turkana (after Lokuruka 2006)

Biological or common name of meat portion/organ	Beneficiary
Head	Male head of awui
Right hind-leg	The male head of awui (claimed by first wife)
Left hind-leg	Female donor of animals
Right fore-leg, normally accompanied by the last four ribs from the right side of the carcass	Male head of family; first wife claims it for him
Left fore-leg	Eldest daughter in the awui
Left pelvic wing	Youngest wife
Oesophagus and windpipe	Youngest wife
Right pelvic wing, usually accompanied by right hind-leg	First wife. May set aside for male head of awui
Neck, thoracic vertebrae and attached muscles	Women of the awui (female neighbours receive donations)
Vertebrae of sacral and coccygeal region including muscles	Paternal grandfather or mother of the head of awui
Sternum meat	Head of awui for distribution to the children
Sternal bones, partly shorn of meat	Female donor of animal for distribution to other wives or female relatives

Turkana of Africa, whereby specific portions of animals are designated to particular individuals so that all receive a cut (Lokuruka 2006). Similar practices were found throughout the medieval period in England, when hunted deer were 'unmade' (butchered in a standardized and ritualized fashion), with different parts of the body being given to particular people (Chapter 8).

Recognizing that different cuts of a carcass may carry higher or lower values as foodstuffs is perhaps not taking the issue far enough. Szynkiewicz (1990) demonstrated how the worldview of pastoral Mongols was structured and performed through their relationship with sheep bones, in particular the 'pure' tibia (which plays a role in the Mongal's origin myth and is incorporated into various rites of passage) and 'devilish' pelvis (which must be kicked on sight). These skeletal elements clearly represented more than just good and bad meat. Similarly, for Malawi, Morris (1998: 141) explains how the claws, skin and horns of dead animals are crucial ingredients in medicine, bringing life to those who consume them (Chapter 6).

These are not rare examples of social meaning being applied to animal body parts; it is more likely to be the rule rather than the exception for the archaeological societies that we study: Choyke's (2010) review of animal-related amulets and their role in power and magic demonstrates just how widespread such practices are (Chapter 6). Davis (2008a) presents a number of case-studies where animal body-part patterns are suggestive of ritual practice, such as Tell Qiri in Israel where five cultic loci produced assemblages dominated by sheep/goat upper forelimbs, notably scapula, humerus and radius. Of the eighty scapulae present, seventy-eight came from the right hand side of the body. Further examples of left and right side preference in animal sacrifice deposits are discussed by MacKinnon (2010a), who concludes that, amongst the Greeks, Romans and Mayans, right-side elements were preferentially deposited in sacrificial contexts, whereas the 'sinister' (literally 'left' from the Latin) specimens were connected to underworld figures.

There is certainly a growing awareness amongst zooarchaeologists that skeletal element and side selection is a real and widespread phenomenon that deserves consideration (Sykes 2007a). As can be seen above, most archaeological investigations have focused on 'ritual' contexts (e.g. temples and shrines) where the patterns are more readily observable. In all probability the significance of animal parts is unlikely to be restricted to ceremonial settings but until such time that zooarchaeologists regularly record and publish all skeletal elements and siding data (many practitioners employ 'rapid' methods that exclude certain bones or assume an equal representation of left and right elements), it seems improbable that these subtle, nuanced patterns will be detected, let alone interpreted.

Ageing

One area that, I feel, has simultaneously advanced and restrained zooarchaeological interpretation is the development of ageing methods, or rather the way in which they are viewed. Standardized recording techniques and experimental cull-models for domestic animals (e.g. Grant 1982; Payne 1973) are great positives. They have transformed understanding of past economies and made it possible to determine whether animals were likely used for primary products (meat, skin, fat, oil), harvested repeatedly for milk, traction or wool, or employed in a mixed economy. However, by focusing on patterns of economic exploitation, other human–animal relationships have been overlooked. This may, in part, be because ageing techniques tend to emphasize the death of animals rather than their life, something clear from the methodological terminology: *kill*-off pattern, *mortality* curve and *cull* profile.

It is easy to forget that even where animals were slaughtered young (for instance Gunter the cockerel was nine months old when we killed him, and Nightshade just eight months), the majority of their associations with humans would have been played out whilst alive, the act of being killed and eaten representing a small, albeit intense, fraction of the total human–animal relationship. In situations where animals lived for years rather than months it has to be expected, as Mlekuž (2007) and Armstrong Oma (2010) have argued, that bonds would have developed between them and the people with whom they dwelt, as is found amongst most modern herding and pastoral societies. Abbink (2003) for instance, details the close affinities between Suri herders of Northeast Africa and their cattle, which are not seen as economic commodities, pets or divine beings but rather as extensions of, and preconditions for, human society – most Suri personal names are derived from cattle coat colour/patterns and all human rites of passage are intertwined with those of cattle. Crate's (2008: 119) observations of the Skaha of northeastern Siberia also highlight the centrality of cattle within human society, that 'cows are like people, they just can't talk'. The Skaha do eventually eat their cattle, as do the Suri, but (as I argued for Gunter) they are valued more in life than death.

For this reason, perhaps we need to reconsider our 'death' profiles as 'life' profiles. By way of example, viewed in terms of human–animal relationships Payne's (1973) model for milk production (Figure 1.4), could equally be read as an anthropocentric profile, where animals are raised primarily for human benefit, dispatched as soon as possible and before they have time to become familiar components of human society. By contrast Payne's wool profiles (Figure 1.4) could reflect a more zoocentric perspective, whereby animals are valued members of the community, allowed to thrive and contribute to society by providing wool, other products or simply companionship. This is not to suggest that the meat/wool labels should be replaced with anthropocentric/zoocentric labels but rather that there should be some recognition that the age profiles of zooarchaeological assemblages reflect human behaviour, experience and worldviews as much as they do economics. Indeed, we might envisage a situation where regional variations in the age profiles of zooarchaeological assemblages (such as those highlighted by Hambleton 1999 for Iron Age Britain) could be interpreted in terms of daily practice and even group identity. For instance the human–animal relationships brought by different management regimes – e.g. pastoralism or farming – can

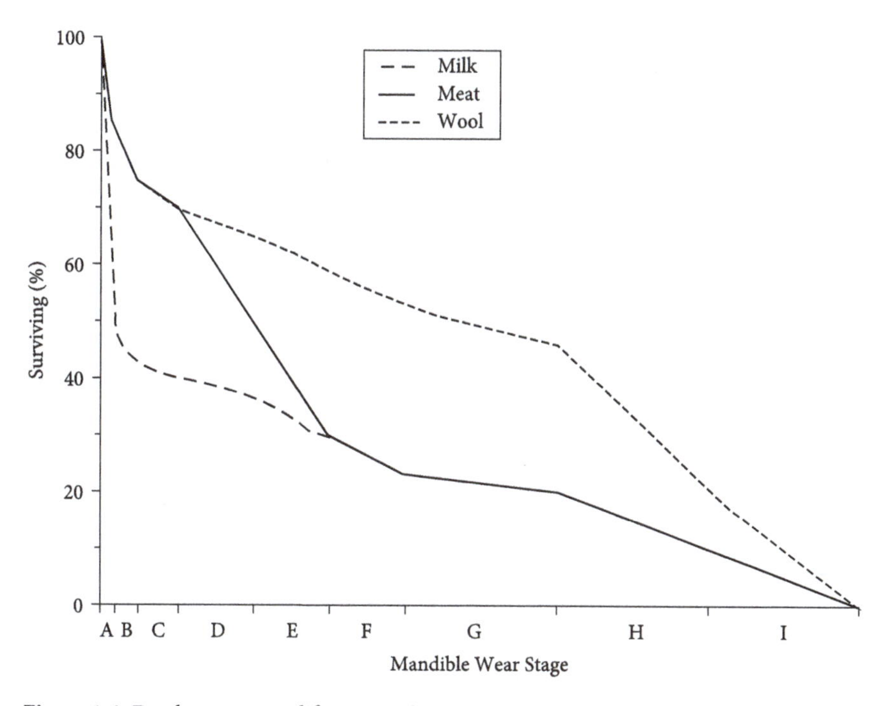

Figure 1.4 Death patterns or life patterns?
Source: after Payne (1973).

produce strong and conflicting group identities (see Shettima and Tar 2008) that could potentially be detectable in the archaeological record through the study of animal age profiles.

Taking this social perspective on animal age further still, I would argue that zooarchaeological age profiles should, wherever possible, be viewed against the available ageing data for contemporary human populations, as comprehension of animal and human mortality are relative and the similarities or differences between the two can be revealing (see Chapter 9). For instance, the death of a 3-year-old animal may be deemed inconsequential to a human in the modern Western world, whose average life expectancy at birth is upwards of 80 years; but how would perspectives change in situations where life expectancy was many decades shorter? An animal life of 3 years may seem a 'good innings' to a people living in a period where the majority of children died shortly after birth (e.g. Bocquet-Appel 2011a, 2011b, highlight the increased infant mortality that accompanied the Neolithic transition). As Gilchrist (2012) has demonstrated clearly, perceptions of life-course and ageing are culturally determined; I would suggest that this is as true for animals as it is for humans.

If zooarchaeologists begin to consider animal age profiles in these terms, ageing data have the potential to tell us as much about human culture and ideology as they do about economies, if not more so.

Sexing

By comparison to ageing techniques, zooarchaeological methods for determining sex are notoriously unreliable, particularly for periods where various breeds of animal are represented in an assemblage (Albarella 1997a). Many techniques have built upon dubious data and, where they have been tested against material of known sex, they have been shown to be widely inaccurate (e.g. Doherty 2013; Sykes and Symmons 2007). It is, therefore, unsurprising that recent zooarchaeological literature lacks confidence in discussions about flock and herd composition. But the sex of an animal *matters*, not only for reconstructing hunting and husbandry regimes (McGrory *et al.* 2012) but also for understanding other human–animal relationships, which must have varied considerably according to the different temperament and physicality of males and females (again, Gunter the cockerel was a very different creature to the hens).

One topic that particularly deserves attention is castration, as this is a cultural act through which biological bodies are transformed and, according to Tani (1996) and Ingold (1994), individuals may be subordinated to the position of slaves. Castration was presumably undertaken for many practical reasons: to make animals easier to handle; for purposes of selective breeding; because the fleeces of wethers are heavier and finer than those of ewes or rams, and also to help the animal produce fatter and tastier meat (Thomas 1983: 93). However, castration events can play an important role in human society. For instance Taylor (2000: 166–9) sees the emergence of animal castration as a pivotal moment in the history of social development, representing a shift in worldview that also gave rise to the concept of human castration: for this reason the origins of animal castration can, to some extent, be seen as a proxy for the origins of eunuchs. However, he has also highlighted our ignorance about the circumstances of ancient castration practices, stating that 'until zooarchaeologists do the work we can only speculate'. New methods are now being developed to enable zooarchaeologists to move beyond speculation (Popkin *et al.* 2012) but, if social zooarchaeology is to prosper, it is essential that these, and other new zooarchaeological techniques, are established with the purpose of answering mainstream cultural, as opposed to purely economic, questions.

Osteometrics

I have less to say on the issue of measurement. It is a necessary and, thanks to von den Driesch (1976), increasingly standardized aspect of zooarchaeology that provides vital information about species identification, the sexual composition of assemblages, and variations in animal size and conformation, the latter most often utilized in studies of domestication and selective breeding (Chapter 2). There have been numerous studies of animal size change, particularly in relation to issues of 'Romanization' in Europe (Albarella *et al.* 2008; MacKinnon 2010b), all of which have identified that significant increases in animal size coincided with periods of Roman occupation. As MacKinnon (2010b: 70) has outlined, it has proved difficult to pinpoint any single reason for this apparent 'improvement' of animal stock, which was more probably multi-causal. However, MacKinnon (2010b: 71) does stress that 'people develop livestock' and he suggests that their decision was probably motivated by increased demand for agricultural foodstuffs and thus stronger, meatier and more productive livestock (MacKinnon 2010b: 70). I have no objection to this economic interpretation but, returning to the mirrors/windows device, I do wonder what the situation means in broader cultural terms: why does improvement happen in the Roman period and what does it tell us about the people responsible? To me, the situation is indicative of an anthropocentric worldview, where animals are seen as a product that can be manipulated by people according to their own desires. This issue of animal commodification will be examined in Chapter 2 but it is also something that can be considered through butchery analysis.

Cut Mark and Butchery Analysis

Unlike measurements, there is very little standardization in the recording of cut marks, although serious attempts are now being made to address this situation (e.g. Orton 2010a; Popkin 2005). Because of this lack of standardization, studies of carcass processing have been somewhat neglected despite the well-recognized potential of cut mark analysis to inform on issues of economic transition, cuisine and ultimately cultural behaviour (see Seetah 2008 for an overview). Grant (1987) was one of the first zooarchaeologists to undertake a comprehensive review of butchery practices in England. She demonstrated that Iron Age and Anglo-Saxon patterns consisted predominantly of knife marks and appeared to be more haphazard than those of the Roman and later medieval periods when butchery practices become more consistent and employ greater use of the cleaver, most probably reflecting the emergence of professional butchers (Chapter 8). Recent studies by Maltby (2007) and Seetah (2007) have largely confirmed Grant's conclusions, that consistency in cut mark patterns and the use of meat cleavers are associated with the rise of professional butchers in the urban environment. Maltby (2007: 72) states that the move towards commercial provisioning would have severely disrupted traditional cultural practices associated with the breaking and redistribution of animal carcasses. However, I would suggest that professional butchery and market redistribution would have completely overhauled human–animal relationships, and it is this change in relationship that I perceive in the butchery marks.

When we killed and butchered our cockerels Gunter and Nightshade, we no doubt left slightly different traces on their skeletons: the two events were different reflecting the different relationships that we had with these two individuals. In both cases the butchery was done carefully and respectfully because we actually liked our cockerels. For this reason, we did not pick up a cleaver and rapidly chop them into evenly-sided potions; aside from the fact it would have caused bone splinters, it would have felt morally inappropriate. No such consideration was afforded the animals that made their way to markets in Roman and later medieval

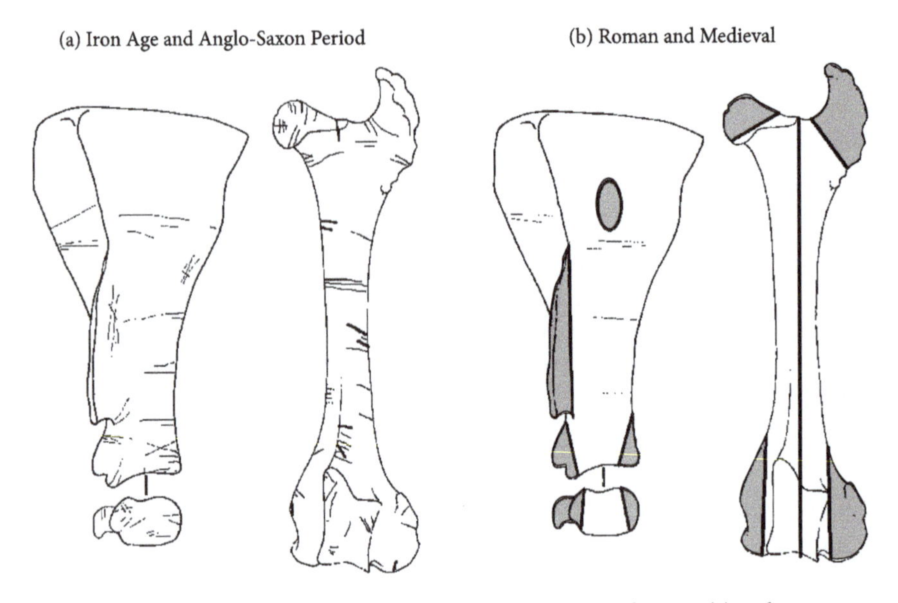

Figure 1.5 Butchery marks typical of Iron Age or Anglo-Saxon deposits (a) and Roman or Medieval deposits (b).

Source: Sykes.

towns. Butchers presumably had little time to build a relationship with the animals they dispatched and any hint of character could quickly be erased by transforming the individual into anonymous and standardized cuts of meat: chop, chop, chop. The Roman and later medieval butchery marks shown in Figure 1.5b scream of a situation where animals are viewed as commodities rather than individuals, something that is less apparent from the Iron Age and Anglo-Saxon butchery patterns (Figure 1.5a) and, I would suggest, this tells us something about the closer human–animal relationships experienced in these periods.

Palaeopathology

Evidence for human–animal relationships can often be found through studies of palaeopathology, skeletal markers of disease and trauma that have the potential to tell us about standards of animal care or levels of animal abuse. Originally, studies of animal palaeopathology tended to focus on individual 'weird' specimens (Baker and Brothwell 1980) but increasingly researchers have sought to undertake species-specific studies at the site level (e.g. see papers in Miklíková and Thomas 2008), between settlements (Vann 2008), across entire regions (Bartosiewicz 2008) or even at the cultural scale, such as is presented in MacKinnon's (2010c) analysis of dog health in the Roman world. As recording techniques develop and become increasingly standardized (De Cupere *et al.* 2000; Vann and Thomas 2006), there is a very real possibility that studies of palaeopathology may provide the clearest insights about attitudes to animals in the past (R. Thomas 2012; Upex and Dobney 2011).

Step 3: Considering Interpretation

As has been outlined above, interpretation should begin the moment we start looking at zooarchaeological assemblages. Yet, we need to realize that animals cannot be the sole focus of our analysis. Human behaviour and thought are not compartmentalized; we do not interact with animals now, plants later and ceramics tomorrow afternoon – our lives are integrated. Similarly, it is important to recognize that the lives of animals are also not compartmentalized; the bones that we study, although recovered from their final resting place, have a back history, or 'biography', that documents a lifetime of interactions with cultural landscape and environment as well as people. How then, can zooarchaeologists ever hope to disentangle such complexity? The answer is that we cannot, or at least not on the basis of bones alone. If it is accepted that the lives of humans and animals are integrated, zooarchaeological studies should be likewise and we must cast our net wide if we are to understand past societies. This is not simply in terms of examining animal bone data alongside other sources of archaeological data, although this should be a priority as evidence from material studies and archaeological science (e.g. ancient DNA, stable isotope and lipid analysis) are invaluable, as is discussed below, but also by considering discussion from other disciplines.

Sociology and anthropology have much to offer us, as our 'Western' society is probably the least well placed to understand the practices and thoughts of those whom we study – our lifestyle and worldview are simply too far removed. I must stress that I am not advocating we that we read anthropological literature, cherry-pick the customs from one culture and then drop them wholesale and uncritically onto another; this is the worst possible practice. However, as a mind-stretching exercise, the consideration of evidence from other societies, as well as information derived from our own personal experiences (e.g. Fairnell 2008), can be exceptionally enlightening.

It is not always necessary to look to other societies for answers however; sometimes the very information that we seek is readily available to us but sits ignored. This is particularly the case for historic societies. Archaeologists often go to great lengths devising complex theoretical models about social practice (often developed from anthropological ideas) without considering the evidence provided by contemporary texts. For instance, throughout much of Europe and for much of the last 2,500 years (probably longer), every aspect of human life – diet, farming practices, health, medicine, behaviour, even life-cycles themselves – was perceived, explained and dictated by elemental theory, or humoural principles (see Arikha 2007; Grant 2000; Jones 2013; Scully 1995). Yet there is little mention of the humours in archaeological discussion; this is an astonishing oversight.

Put simply, elemental/humoural theory viewed all things within the known universe as combinations of the four elements – air, earth, fire and water – and all matter as a mixture of four opposing temperaments: hot, cold, dry and moist. In turn these combinations of elements and temperaments made up the four humours, all of which are found in living bodies:

(1) Sanguine (blood), which corresponded to air and was hot and moist
(2) Phlegm, which corresponded to water and was cold and moist
(3) Choler (yellow bile), corresponded to fire, being hot and dry
(4) Melancholy (black bile), linked to earth which was cold and dry

Humours were not fixed but found in varying quantities in different individuals (including plants and animals) according to the season of the year and the individual's sex, age and personal character (Figure 1.6). Humoural make-up changed through an individual's life – starting moist and warm, becoming progressively dry with maturity until the cold/wet decay of old age – and could be influenced by the temperament of the foods/drinks they consumed, the environment/climate in which the person lived or the company in which they mixed. Temperaments could also be transferred through sight, sound and touch, as well as through consumption. The maintenance of an individual's humoural balance was considered vital, with any deviation having the potential to cause serious damage to health. For instance Hernandez (2010) has shown that iced drinks were widely believed to be lethally 'cold', inspiring medical debate and public health warnings right up to the end of the sixteenth century. Eventually it was accepted that men (whose temperament was deemed relatively hot and dry) might be able to tolerate the cooling effects of such drinks but for women, who were considered to be comparatively cold and wet, iced drinks should be avoided at all costs.

This sixteenth-century controversy over cold-drinking may seem a ridiculous example of Renaissance medicine, a period when blood-letting was also all the rage; however, humoural philosophy was widely held and had a very deep antiquity. It was the cornerstone of medieval thought, influencing the diet and daily practice of all sections of society (Jones 2013; Scully 1995). But this medieval doctrine was founded upon even earlier traditions of Roman and Greek scholars for whom life was equally conceptualized and lived according to humoural principals: Pliny's Natural Histories, particularly the sections on medical properties and veterinary practices, are replete with evidence for humoural thought (Swabe 1999: 73–4). Roman subscription to humoural philosophy is perhaps most famously provided by the works of Galen (*c.* AD 131–216), who himself was transmitting ideas contained in the Hippocratic Corpus of ancient Greece. The earliest written evidence for elemental thinking in Europe has been ascribed to the pre-Socratic Greek philosopher Empedocles whose poem, *Phusis,* written in the fifth century BC clearly lays the foundation for later humoural theory (Arikha 2007; Jones 2013: 12). It is noteworthy that evidence for elemental philosophy is also found in the literature of ancient India; for instance the *Rigveda* (the body of Indian Sanskrit texts composed between 1500–500 BC) contains passages that are very similar to ancient Greek traditions suggesting that both share a common ancestry (Arikha 2007: 303).

Although a clear line of descent can be found for humoural theory in Europe, it is becoming increasingly apparent that, as with systems of animal classification (see

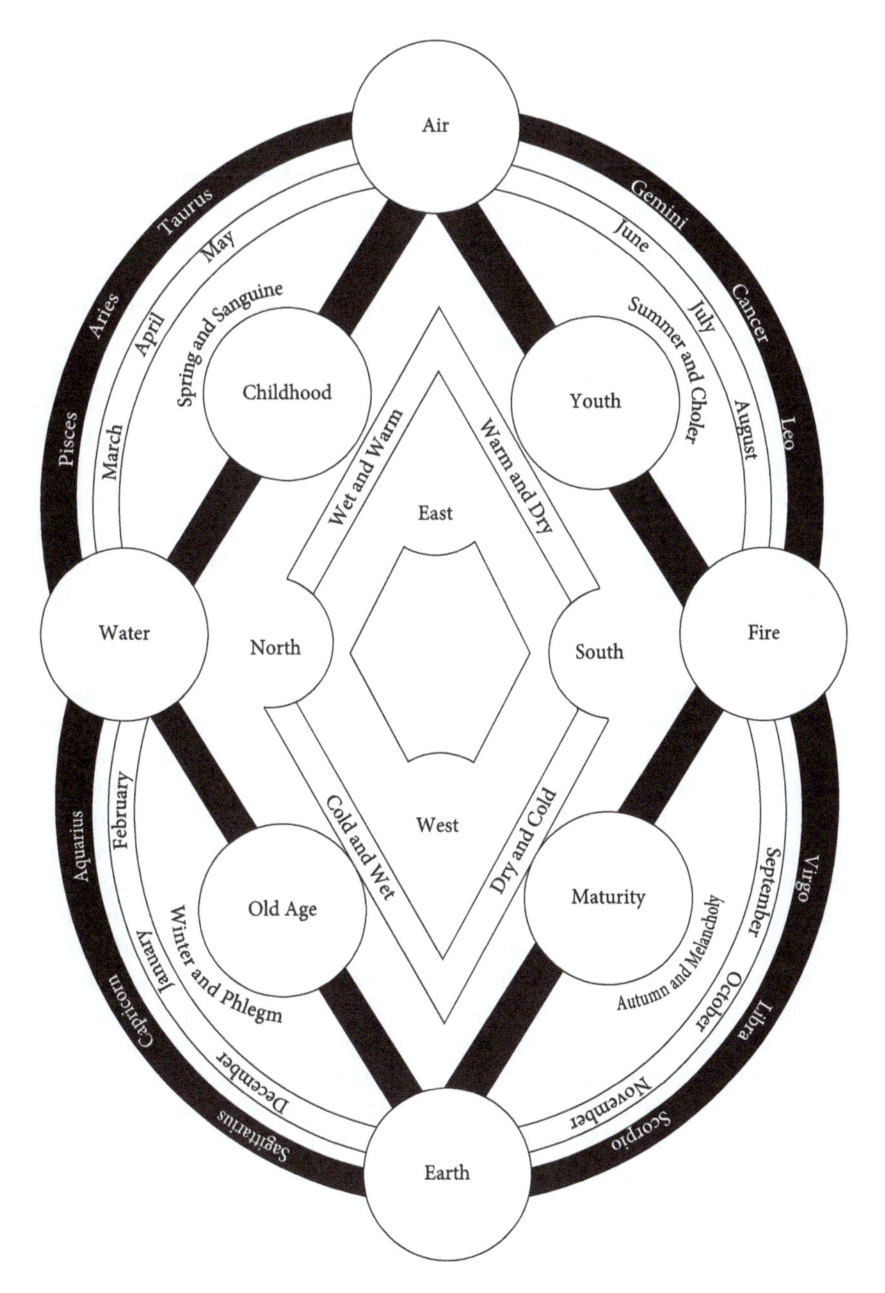

Figure 1.6 Medieval Worldview.

Source: Richard Jones.

the section 'Taxonomy' above), humans have an almost universal tendency to develop elemental principles, and in particular hot/cold classifications (Anderson 1987; Morris 1998, 2000). Whilst extreme variations exist between what is considered 'hot' or 'cold' in any given culture we should probably expect that, throughout the world and throughout most of (Pre)history, human behaviour and thought were most likely governed by similar overarching concepts. This has some implications for our understanding of human–animal relationships because, within most humoural systems, the temperament or character of animals can be acquired by the people who interact with them: animals are considerably more important within humoural philosophies than they are in modern 'Western' thought.

The Enlightenment, with the rise of modern Western science, is often cited as responsible for the demise of humoural principles and for creating the separation between humans and animals that led to the subjugation and objectification of the latter (Bulliet 2005: 45; Ingold 1994; Thomas 1983). This is not a debate that I wish to rehearse here except to say that, to some extent, recent trends in archaeological science have begun to redress the balance. Interactions and elemental transfers between humans, animals and material culture can now be gauged through analyses of biomolecules; for instance studies of animal-derived lipids in pottery have proven the existence of widespread dairying in Prehistory (see Chapter 2). Furthermore, it is now widely recognized that isotope studies of humans are bereft without first undertaking isotope analyses of animals – in common with humoural theory, isotope studies seek to understand, quite literally, how elemental composition is transferred from the landscape and from one organism to another.

The isotope elements most frequently analysed for skeletal remains are the stable isotopes of carbon (^{13}C and ^{12}C) and nitrogen (^{15}N and ^{14}N), which can be analysed for any body tissue (bone, tooth, hair, nail or hoof) to inform on dietary variation between individuals and populations. The ratios of these two sets of isotopes ($\delta^{13}C$ and $\delta^{15}N$ respectively) vary between different ecosystems and the underlying principle of isotope analysis is that the composition of food and drink derived from different ecosystems will transfer through the food chain becoming reflected in the consumer's body tissues (Ambrose and Norr 1993; Tieszen and Fagre 1993). At the most basic level, gross variations in bone collagen $\delta^{13}C$ values are thought to be influenced by the consumption of marine versus terrestrial protein (Schoeninger *et al.* 1983) or by the inclusion of arid C4 plants in the diet (Vogel and Van der Merwe 1977); whereas $\delta^{15}N$ values reflect the proportion of plant and animal protein in the diet, becoming more enriched at each trophic level (Ambrose and Norr 1993; Richards and Hedges 1999). These broad variations are shown in Figure 1.7, which provides a typical stable isotope 'map' for different ecosystems.

At a more detailed level researchers are increasingly highlighting the complex range of variables – e.g. temperature (Stevens *et al.* 2006), water availability (Schwarcz *et al.* 1999), salinity and marine input (Guy *et al.* 1986a; 1986b; van Groenigen and van Kessel 2002; Britton *et al.* 2008) – that can result in small-scale variations in the isotope composition of plants and animals, so influencing the values for humans. For this reason isotope studies of animals are considered fundamental for the interpretation of human values. As a result there now exist large isotope datasets for

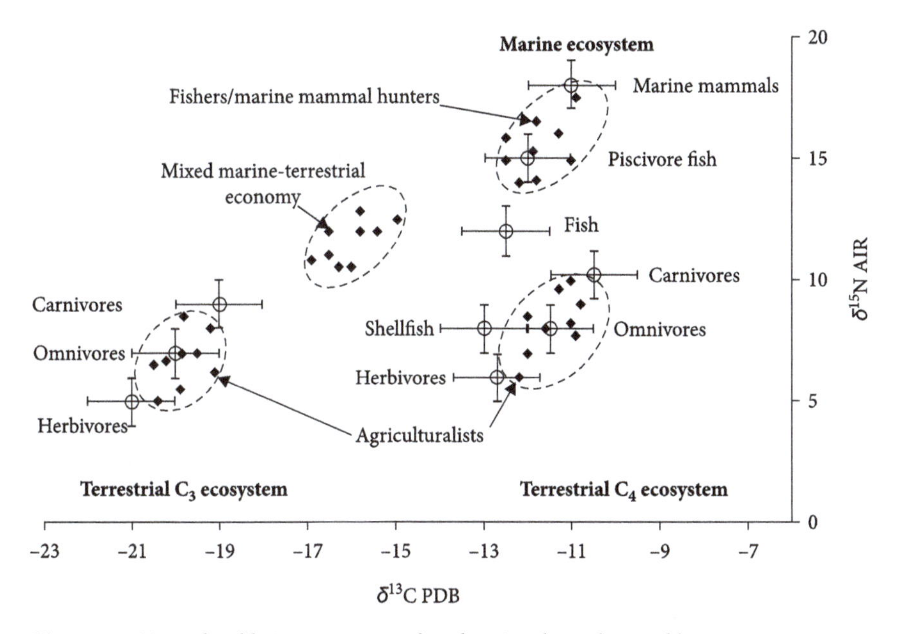

Figure 1.7 Typical stable isotope scatterplot, showing the carbon stable isotope ratios on the x-axis and the nitrogen stable isotope ratios on the y-axis.

Source: Holly Miller.

archaeological animals; however, few studies consider animals in their own right – they are generally viewed simply as background data for research into human diet. This would seem to be a missed opportunity because close examination of the animal data can reveal important social evidence for human–animal relationships, informing on issues from domestication and management (Barton *et al.* 2009; Rosvold *et al.* 2010; Madgwick *et al.* 2013 – see Chapter 2), pet-keeping (MacKinnon 2010c – see Chapter 7) and animal mobility (Pearson *et al.* 2007 – see Chapter 2).

Whilst issues of movement and migration can be addressed at low resolution through $\delta^{13}C$ and $\delta^{15}N$ analyses, the ratios between the isotopes of strontium ($^{87}Sr/^{86}Sr$) and oxygen isotopes ($\delta^{18}O/^{16}O$) provide more reliable information relating to translocations. As with carbon and nitrogen, both sets of isotopes transfer through the food chain leaving a signature in skeletal tissue – the former relating to geological terrain, the latter to climatic variations in drinking water. Unlike $\delta^{13}C$ and $\delta^{15}N$ analysis, however, not all skeletal materials are suitable for analysis. For archaeological studies, tooth enamel is the preferred material because, unlike bone, it is resistant to diagenetic change and, since enamel is not remodelled throughout life, the continual growth structure of teeth preserves strontium and oxygen composition allowing seasonal changes in human and animal movement and diet to be reconstructed.

Using these scientific techniques we are able, more than ever before, to tell the stories of individual animals. For instance Bendrey *et al.* (2009) utilized strontium isotope analysis to contrast the lives of two Iron Age horses – one bred locally in southern England, the other imported from further afield. Similarly Berger *et al.* (2010) used both strontium and oxygen isotope analyses to reconstruct the life history of a mule recovered from the Roman fort in Bavaria. These intimate life histories are known only because of advances in archaeological science and, as the cost of analysis reduces, isotope investigations have the potential to transform zooarchaeological research.

Other forms of scientific investigations – such as genetic analyses – also have the capacity to provide new perspectives on old questions. For instance, Pruvost *et al.*'s (2011) genetic analysis of pre-domestic horse bones from across Europe and Siberia was able to revitalize debates about Paleolithic cave art. Through DNA analysis Pruvost *et al.* (2011) isolated the genetic markers that determine horse coat colour/pattern and, when the same marker was examined in archaeological specimens they were able to identify Palaeolithic individuals with bay, black and 'leopard' spotted coats, all of which are depicted in Palaeolithic cave art. Genetic identification of spotted horses is particularly important in this context because representations of spotted animals have, in the past, been interpreted in terms of 'hyperimagery', as creatures imagined during trances. By taking an integrated approach it has been possible to challenge traditional interpretations and suggest that the cave art is depicting real rather than imagined animals.

Artistic representations of animals, both static and portable, provide important information about how human societies conceived and understood animals (e.g. Kalof 2007; Morphy 1989). However, as with all sources of evidence, visual culture is not straightforward; it is notoriously difficult to decipher, especially in the case of Prehistoric art (Bahn and Vertut 1988; Morphy 1989). Identifying the kinds of animals represented in works of art is frequently problematic and interpreting the meaning of the images is often impossible, especially if the evidence is not contextualized. Zooarchaeology has the capacity to provide such contextualization and Bahn and Vertut (1988) provide a variety of examples to show that where Paleolithic art has been compared with local contemporary animal bone assemblages, the two seldom correspond: for instance reindeer comprise 90 per cent of the zooarchaeological assemblage from Lascaux in the French Dordogne, 75 per cent of that from Comarque, Dordogne, and over a third of the material from Gata de Gorgos in Spain, but at all of these sites depictions of reindeer are either rare (e.g. there is one reindeer image at Lascaux) or completely absent (Bahn and Vertut 1988: 177–9). Such contrasts in evidence do not provide answers to all our questions but they do highlight the necessity of integrating different datasets if we are to develop anything more than the most simplistic interpretations of past societies.

Uniting evidence from a wide range of sources is, in my opinion, vital but it does run the risk of turning any study, however minor, into a work of unwieldy magnitude. It is, therefore, important to be realistic. Time and money dictate that not every zooarchaeological report can be ground-breaking but our ultimate aim must always be to advance knowledge rather than simply repeat it. This is not any easy challenge to set ourselves, but it has the potential to be rewarding and thought-provoking, as I hope the next few chapters demonstrate.

Chapter 2

ANIMAL 'REVOLUTIONS'

Zooarchaeology has built its reputation on the study of revolutions: the 'Neolithic Revolution' when people first domesticated animals and commenced an agricultural lifestyle; the 'Secondary Products Revolution', purportedly a Bronze Age phenomenon that heralded the use of animals for milk, wool or traction; and, for historic Europe, the 'Agricultural Revolution', when selective breeding of animals gave rise to new and 'improved' breeds of animals. All three have given animal bone studies the rare opportunity to sit at the centre of archaeological debates and zooarchaeologists have been instrumental in demonstrating that these 'Revolutions' were nothing of the sort, at least from a temporal perspective. Far from overnight transformations each was characterized by complex changes that occurred over long (but varying) periods of time and with outcomes that differed by location.

Nobody would deny that these 'revolutions' represent important shifts in the (Pre)history of human–animal relations: the transition from hunting to animal husbandry, the decision to utilize animals in life rather than just death, and the move towards intensive animal breeding all reflect very different forms of engagement. However, many zooarchaeological studies have focused more on identifying patterns – in particular changes in animal size, shape and demographics – that might feasibly be taken as evidence (better still the earliest evidence) for a revolution, rather than what these patterns might mean in terms of human daily practice, experience and ideology. In this chapter I will reconsider the evidence and offer some alternative musing on these three so-called revolutions.

Domestication and the Neolithic Revolution

No other topic in zooarchaeology has attracted more attention than domestication. The literature surrounding the issue is expansive, ever growing, fast changing and cannot be summarized here in anything but the most general terms. For me, this is fortunate, as my research has never focused on early Prehistory so, academically, I am not qualified to comment on the subject (those with an interest in domestication should turn to Russell 2002; Vigne 2011; Zeder 2011a, 2011b, for excellent syntheses and new perspectives on the literature). Indeed, my lack of knowledge is such that

I must start this chapter with a confession: I do not really understand what the term 'domestication' means, it seems to be a catchall for everything and yet nothing. Almost without exception, publications on domestication start with a review of all the various definitions. Some are devoted entirely to reviewing those definitions (O'Connor 1997), whilst others state that it is a 'necessity' to first define what is meant by domestication before proceeding to discuss it (e.g. Arbuckle 2005: 19; Orton 2010b: 189; Vigne 2011: 172). Because of this apparent necessity, much ink has been spilt over the subject but, after more than 50 years of continuous debate, the term domestication remains ill defined – so perhaps I am not completely alone in my confusion. In keeping with tradition, I will outline a few of the different definitions here, although Russell (2012: 207–58) and Zeder (2011a, 2011b) should be consulted for the most comprehensive and up-to-date reviews of the debate.

Defining Domestication

Most, but not all, scholars would agree that domestication was not an event but a process, involving different human–animal relationships at different times in different places: from (a) loose contact, whereby humans co-existed with and even tamed wild animals but did not necessarily influence their breeding, to (b) the keeping of animals whereby the movement and breeding of herds was restricted, and eventually to (c) complete human control of entire populations with implications for animal nutrition, genetics, physiology and behaviour. At what point along this continuum, and by what mechanisms, the status of animals changed from 'wild' to 'domestic' is debatable and it is here that the controversy begins.

Perhaps the most common understanding of domestication is that it relates emphatically to the long-term control of animals for human profit and, as such, human intentionality is the determining factor, with animals being passive subjects in the process (e.g. Bökönyi 1969: 219; Clutton-Brock 1994: 26–30; Meadow 1989). Other researchers, however, (e.g. Budiansky 1992; O'Connor 1997; Zeuner 1963) have stressed that humans do not always exert all the control; that animals arguably domesticated humans through a co-evolutionary process of mutualistic symbiosis. Certainly animals have the capacity to alter human behaviour and the relationship between the two can often be viewed as a partnership, albeit one where animals are the junior partner with most of the life or death decisions being made by humans. For Ingold (1994) and Russell (2002) it is these very power relations, notably the appropriation of animals as property but also the potential of animal ownership to alter human–human relationships, which both characterize domestication and render it such an important area for study.

Although scholars such as O'Connor (1997), Ingold (1994) and Russell (2002) have provided new ways of thinking about domestication, they do not offer new definitions. This is probably the best way to proceed because most attempts to do so have resulted in classifications that are either so nebulous that they are meaningless or so restricted that they exclude a myriad of human–animal relationships that are included within other definitions of the term (O'Connor 2000: 149). For instance peregrine falcons are generally considered to be 'wild' animals, even when they are enclosed in zoos, but what about when they are tamed and bred for use as hunting aids (Oggins 2004)? Similarly, dogs and ferrets are widely perceived as domestic animals but what about when they turn on their owners (Paisley and Lauer 1988)? There is perhaps a need to examine the evidence on a case-by-case basis.

Even if we were to agree on a broad definition of domestication, such as the conciliatory version provided by Zeder (2011b: 164):

> A sustained, multigenerational, mutualistic relationship in which humans assume some significant level of control over the reproduction and care of a plant/animal in order to secure a more predictable supply of a resource of interest and by which the plant/animal is able to increase its reproductive success over individuals not participating in this relationship, thereby enhancing the fitness of both humans and target domesticate.

the identification of early domesticates in the archaeological record is no easier, especially for those regions of the world where 'domestic' animals first emerged from their wild progenitors.

Identifying Domestication

Up until recently, it was assumed that wild and domestic relatives could be separated on the basis of skeletal size and conformation: domestic individuals being markedly smaller and retaining more juvenile features than their wild progenitors (Davis 1987: 126; Grigson 1969). Certainly diminution in animal size has been shown repeatedly in Neolithic assemblages, indicating that the phenomenon is real and linked to various factors (e.g. selective breeding, nutrition and health) associated with the domestication process (Helmer *et al.* 2005). However, although studies of modern animals have demonstrated that size and shape change can occur rapidly when populations are taken into captivity (e.g. Arbuckle 2005; O'Regan and Kitchener 2005) this is not always so. Indeed, zooarchaeological research is beginning to show that, in the majority of cases, the origins of animal management pre-dated the appearance of detectable morphological change by many centuries (Zeder 2005). The same issue applies to genetic variation; for although DNA studies, particularly phylogeography, are powerful tools for investigating the geographical origins and spread of domestic animals (e.g. Zhang *et al.* 2013) it is over-optimistic to expect that they can pinpoint the moment that human–animal relationships first changed. It is true that genetic studies of coat colour are very promising in this respect, as modern experiments have shown that the influence of human-imposed selective breeding can have a rapid influence on pelage patterns (Trut 1999). However these variations, like size and shape change, do not happen overnight; they occur as a part of the domestication process and take time (Larson 2011).

Understanding the earliest origins of domestication may require that greater attention is given to sources of evidence beyond animal bones, especially if Hodder

(1990) and Cauvin (2000) are correct in their suggestions that the process of domestication was originally symbolic or ideological rather than physical. Under such circumstances, wider shifts in material culture, artistic representation or human *treatment* of animal remains (rather than the animal remains themselves) may be the best indicators that animals had crossed the boundary from wild to domestic. That said, it would be a sad indictment of the profession if zooarchaeologists were completely unable to detect the transition from hunting to farming.

Over the last decade zooarchaeologists working on Epipalaeolithic and Early Neolithic assemblages from the Near East have made considerable progress refining their techniques of analyses, and demographic profiling has emerged as one of the most sensitive gauges for detecting the new human–animal relationships brought by domestication. The premise of demographic profiling is that the priorities of hunters and herders would be sufficiently different to generate distinctive age/sex structures in archaeological assemblages: hunters would employ more of an 'optimal foraging strategy' to target animals that will maximize meat return (e.g. large males), whilst herders would aim to maintain the long-term viability of their herd by selectively culling young males but leaving females to breed. Demographic profiling has been applied convincingly by Zeder (2005, 2006) whose study of goat assemblages from the pre-pottery Neolithic sites of Ganj Dareh and Ali Kosh in Iran was able to identify sex-specific age patterns consistent with the hypothetical models for herd management. This early evidence for goat management/domestication is strengthened by the absence of similar patterns in Epipalaeolithic assemblages but also in contemporary collections of gazelle remains, which presumably continued to be caught through hunting. Importantly, these sex-specific age profiles became apparent almost 1,000 years before any change in goat skeletal morphology was observed in the zooarchaeological record, reinforcing the idea that size change is a poor indicator of early domestication.

Following Zeder's work, similar demographic approaches have been applied to sites across the Near East, highlighting the regional and temporal variations in the transition from hunting to herding (see Arbuckle 2012 for a review). At the late ninth/early eighth millennium settlement of Aşıklı Höyük in Cappadocia, for example, the demographic profiles for both sheep and goat show early evidence for herd management, with a clear emphasis on the culling of animals aged between 1 and 3 years of age. The sex distributions for sheep and goat varied however – the profile for Aşıklı Höyük sheep having a slight bias toward the culling of females, whereas the goat assemblage contained more males – suggesting that the two taxa were managed in different ways (Arbuckle *et al.* 2009; Buitenhuis 1997). As with Zeder's (2005) study, this selective pattern occurs without the animals displaying any hint of morphological change, leading Buitenhuis (1997) to conclude that the animals were managed with a light touch rather than being taken into physical captivity. Based on the animal bone data alone, Buitenhuis's suggestion would seem the most likely scenario; however, the assemblage from Aşıklı Höyük is one of very few that has been subject to stable isotope analysis (Pearson *et al.* 2007) and the results are not what might have been predicted.

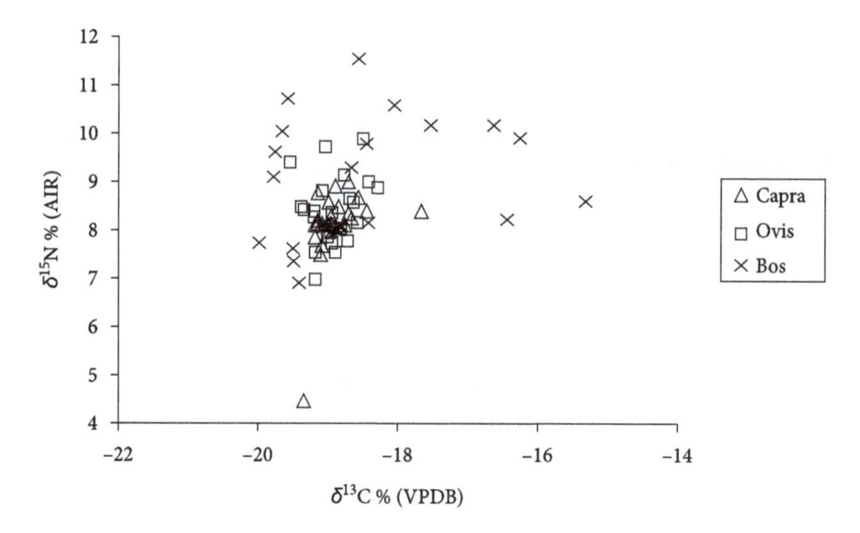

Figure 2.1 Capra, Ovis and Bos stable carbon ($\delta^{13}C$) and nitrogen ($\delta^{15}N$) isotope values of bone collagen from Aşıklı Höyük – note that the distribution of the goat and sheep results is far more restricted than that for cattle, which are assumed to be wild.

Source: Jessica Pearson.

Figure 2.1 shows the stable carbon and nitrogen isotope data for Aşıklı Höyük's sheep and goats. It can be seen that both taxa cluster tightly together, especially by comparison to the diverse range of measurement for Aşıklı Höyük's wild cattle. Pearson *et al.* (2007) interpreted the data as evidence that humans were beginning to take local caprine populations into captivity, restricting their movement and thus their diet. Such a scenario is at odds with Buitenhuis's (1997) interpretation, and Arbuckle *et al.*'s review of all the evidence for the site concluded that the nature of caprine exploitation at Aşıklı Höyük was 'unclear' (Arbuckle *et al.* 2009: 147–8, 150).

Certainly the different sets of data for Aşıklı Höyük do appear incompatible but they are, nevertheless, both valid and therefore demand interpretation (my suggested interpretation can be found on page 31). Indeed, it is often the integration of apparently contradictory sources of evidence that produces the clearest understanding of the human–animal relationships that they reflect. We can be confident that these relationships are reflected everywhere in the archaeological record – the difficulty is extrapolating and understanding them. To do so, it is necessary to weave together evidence from different disciplines, scales and perspectives: the more the better. As outlined in Chapter 1 (page 20) isotopic analysis is one of the most valuable sources of evidence for unpicking human–animal relationships, particularly those linked to domestication (Barton *et al.* 2009; Rowley-Conwy *et al.* 2012); however, as the technique is expensive, its application continues to be limited in zooarchaeological studies.

Palaeopathological studies are far more cost efficient and recent advances in methods of description, recording and analysis (Vann and Thomas 2006) are beginning to show the great potential of palaeopathology for providing important information about shifts in the treatment of animals that accompanied the earliest stages of domestication (Upex and Dobney 2011: 197). It is not enough to rely solely on palaeopathological evidence and, wherever possible, multiple strands should be brought together. For instance, in their study of the pigs from Çayönü Tepesi in south-eastern Anatolia, Ervynck *et al.* (2001) found that rates of enamel defects linked to physiological stress (linear enamel hypoplasia) gradually increased as age profiles shifted towards a pattern typical of herding (the slaughter of unnecessary males). These changes were accompanied by a reduction in the length of molar and tooth rows, only later followed by an overall reduction in animal size. Taken together these multiple strands of evidence strongly support the long-held belief that the region around Çayönü Tepesi was one of the earliest centres of pig domestication and that, rather than an overnight transformation in human–animal relationships, the process was slow and gradual.

Ability to detect domestication zooarchaeologically is fundamental to understanding the timing and pace of the process but, as ever, the ultimate aim of zooarchaeological research should be to interpret patterns, not just highlight them.

The How and Why of Domestication?

As advocated in Chapter 1, we might come closer to understanding the circumstance and meaning of the shift from hunting to herding if we pay more attention to the living animals and consider how human interactions with them are likely to reflect wider human society and thought. Keeping in mind the concept that human–animal relationships are a reflection of human society, it is interesting to note Leach's (2003) observation that many of the criteria used to identify the presence of domestic animals - notably diminution in body size and changes in cranial morphology and dental features - can equally be applied to near contemporary human populations which, Leach claims, show parallel changes. To Leach's observations can now be added Bocquet-Appel's (2011a, 2011b) work on human demography, which has demonstrated that, regardless of geographical location, the Mesolithic-Neolithic transition was accompanied by a substantial increase (typically 20 to 30 per cent) in the proportion of immature skeletons in human cemeteries. As has already been demonstrated, the same demographic trait is one of the signifiers that a zooarchaeological assemblage reflects herding rather than hunting. Furthermore, numerous studies of human palaeopathology have highlighted that, as has been demonstrated widely for animals, there is an increase in tooth enamel hypoplasia and other palaeopathologies concurrent with the transition to intensified agriculture (Upex and Dobney 2011: 197). These trends are perhaps just a coincidence; certainly Zeder's reply to Leach's argument (see Leach 2003: 363) stresses that the mechanisms behind the changes seen in humans and animals are likely to have been very different. However, I feel that the evidence is compelling enough that we should no longer examine the (Pre)history of humans and animals separately – they are inextricably linked and ought to be considered in tandem if we are to understand either individually.

Shipman (2010) gives animals a greater role in the human story, putting them at the very centre of all major steps in human evolution – from tool use, to the development of art, religion and domestication. She makes a case for what she terms 'the animal connection'; her hypothesis is that, throughout time, humans and animals have been intimately and persistently connected, and that human adaptive changes were the direct result of this connection. Shipman argues that intensive observation, handling and intimacy with animals, as evidenced by the detailed artistic representations seen in cave paintings, were the pre-requisites for animal domestication. In many respects I agree with Shipman and I like the emphasis she places upon animals; however, she views animals in passive terms, suggesting that they were essentially 'tools' engaged by humans to give them a selective advantage. Whilst this may have been the case, I prefer to envisage animals as more active players in the process of human–animal domestication and am inclined more towards symbiotic and mutualistic models proposed by the likes of O'Connor (1997, 2013b) which allow animals a degree of agency.

Most authors now agree that cats, dogs, chickens and possibly pigs initiated their own domestication, perhaps initially scavenging at the edge of human occupations before moving into mutually tolerated and eventually mutually agreeable relationships with people. This is something that Zeder (2011a, 2011b) has labelled the 'Commensal Pathway' to domestication. However, such a process is widely perceived as unlikely for other common domesticates, Russell (2012: 217) suggesting that the move from hunting to herding taxa such as cattle, sheep and goats would have been too big a leap in terms of lifestyle and logistics to have occurred by accident and without human intent. I accept that an abrupt change from hunting to herding might cause people to look around in puzzlement but, as has been outlined above, domestication was a process, and I believe that it is possible for individuals to enter into new and life-changing relationships without being aware of the transformation occurring. To take a very personal example, I cannot remember or pinpoint the moment that my partner and I went from being two free and independent individuals to becoming completely inter-dependent: was it on the first date (taming)? The first kiss (petting)? When we established a joint bank account (milking)? When we moved in together (domestication)? Or when we took the decision to have children (selective breeding)? All I know is that, 15 years down the line, it would create chaos if we now attempted to extricate ourselves from the partnership: too much has changed and I did not notice any of it happening.

To some extent Zeder (2011a, 2011b) allows for this kind of unconscious fumbling route to domestication with her 'Prey Pathway' model, arguing that, at some point, human hunters who sought to increase their success employed game management strategies that gradually turned into herd management and eventually controlled breeding. Zeder (2011a: 242) does not see the animals as being the instigators for this shift, suggesting instead that it was perhaps a human response to the depletion of local stock. Vigne (2011: 179) suggests that, in all probability, the origins of domestication were multi-causal: a mixture of climate, environment and human biogeography, demography, techno-economic practices, diet and health, social structure and mentality. This is no doubt the case but, once again, it is notable that the animals themselves do not feature in his list of potential causes.

To me, it seems eminently possible that, given the behaviour of hunter-gatherer populations is closely associated with that of the animals they hunt, the changes that occurred amongst some hunter-gatherer groups (e.g. a shift towards sedentism) could have been dictated by the animals themselves (e.g. a reduction in migration). As early as the 1970s Noe-Nygaard (1974) hinted that this may have been a possibility, based on her analyses of hunting injuries in Danish Mesolithic assemblages. She found that in the earlier phases of the Mesolithic most of the animals that displayed hunting wounds were killed outright by their injuries (70 per cent of the injuries were apparently fatal). By the later phases of the Mesolithic, however, a much higher proportion displayed not only fatal wounds but also healed injuries: 87 per cent of those specimens with hunting wounds had been shot on previous occasions but had survived, in other words, just 13 per cent of animals were killed on the first attempt. It seems highly unlikely that the pattern can be attributed to inter-period variations in hunting skill (i.e. that later Mesolithic populations were just a bad shot) and Noe-Nygaard (1974: 245) proposed that the rise in healed injuries may suggest that through the course of the Mesolithic both animals and people were becoming increasingly sedentary and so more likely to meet each other on multiple occasions. Noe-Nygaard (1974: 245) was cautious in her interpretations, warning against issues of sample bias; however, recently her thesis has been revisited by Leduc (2012), who has found additional evidence to support Noe-Nygaard's original hypothesis: it appears that, at least in some areas of northern Europe, animals and the Mesolithic people that relied upon them, became sedentary together.

Such a situation should come as no great surprise as behavioural studies of modern animals have demonstrated repeatedly that migratory species will quickly become sedentary where food supply is sufficient, whether this is provided by the local environment (Martin 2000) or by humans in the form of supplementary fodder (Fletcher 2011: 28; Kothari and Sharma 2013; Mysterud 2010). Indeed, anthropological work by Kothari and Sharma (2013) has shown that the villages of the Bishnoi community in Rajasthan are 'swarming' with antelopes, gazelle and other wild animals, which live amongst the human community because the people provide them with food on a daily basis. With this evidence in mind, it seems plausible that the climatic amelioration of the Holocene, combined with rising temperature and rainfall (Gupta 2004), would have transformed the growth and diversity of vegetation enabling animal populations, and thus human groups, to become increasingly sedentary. There is also some evidence (Tolan-Smith 2008: 148) to suggest that Mesolithic populations were collecting 'leafy hay' to providing additional fodder to animals. Under such circumstances humans and animals could have slipped, without much conscious intent, into new kinds of familiar relationships. Modern animal studies have also shown that, as wild animal populations become habituated to humans they can lose their fear response, even

Figure 2.2 Wild red deer in the Scottish Highlands are easily attracted to forage during the winter. Here the keeper can approach easily, simply by feeding them hay.
Source: John Fletcher.

when they are hunted by people (Mysterud 2010: 921), allowing even closer associations to form (Figure 2.2).

A situation of increased animal sedentism, perhaps first encouraged by climatic change and then perpetuated by supplementary feeding, would certainly help to explain why animal domestication occurred repeatedly around the world in multiple centres, at different times but under similar climatic conditions (Vigne 2011: 179). Such a situation could even account for the apparently paradoxical evidence from the early Neolithic site of Aşıklı Höyük, Anatolia (see page 27), where the isotopic data indicate that caprine mobility and diet were restricted but the remains of the animals themselves appeared closer in morphology to wild than domestic animals: it seems possible that the evidence reflects supplementary feeding and management of local sheep and goats that *chose* to form closer relationships with people but were not *forced* to do so. I am not suggesting this as a new universal model for animal domestication, largely because it is not a new suggestion, but the idea that animals were as responsible as humans for the origins of domestication seems at least as valid as any other explanation, because it allows intentionality and adaptability on the part of the animals. The hypothesis that animal sedentism encouraged human sedentism would be easy to test through a comprehensive programme of isotopic analysis. As has already been shown for Aşıklı Höyük, carbon and nitrogen isotope analysis can be useful in this respect but they are not the optimal methods for ascertaining patterns of animal movement; better indications can be obtained from inter-tooth analyses of strontium and oxygen isotope analyses (see Chapter 1, page 21). Excellent work has been carried out by Britton *et al.* (2009, 2011) who have shown, by studying modern populations of caribou and bison and comparing the results to analyses of Palaeolithic specimens of the same species, that migration patterns can be identified archaeologically. There would seem to be great potential for applying similar methodologies to questions of domestication to pinpoint when, or whether, free-ranging wild animal populations became sedentary as a result of increased contact with humans. As yet, this potential has not been realized beyond a few baseline studies of modern material (Mashkour *et al.* 2002; Wiedemann *et al.* 1996) and small-scale analyses of Neolithic animals to determine the origins of transhumance (e.g. Henton *et al.*'s 2010 work on sheep from Çatalhöyük in Turkey).

Whatever pathway animals took to domestication, at some point the relationship between humans and animals did reach a point that was very different to the interactions that had preceded it, with dramatic implications for all concerned.

The Meaning of Domestication

In his seminal article, Ingold (1994) contrasted hunter-gatherer perceptions of human–animal relationships and those that characterize fully formed pastoral societies, arguing that domestication brought a fundamental shift in worldview and human–animal relationships. According to Ingold (1994), and other subsequent researchers of hunting societies (e.g. Willerslev 2007) hunter-gatherers, in general, perceive no separation between, or even concept of, culture and nature: humans, animals, plants and other beings are all part of the same inter-connected world and are viewed as equal beings, there is no human-dominated hierarchy. Hunter-gatherers relate to their environments in terms of trust, respect and sharing; in much the same way as a parent provides for a child, the environment provides for the human community so long as the community behaves appropriately (Chapter 3, page 58)

By contrast, Ingold (1994) argues that pastoralist communities have a hierarchical relationship with their environment, with people at the top of the chain-of-being and animals very much below. Whilst people may genuinely care for animals, they perceive them to be human 'property'. This, according to Ingold, represents an abrogation of trust and a denial of the animals' autonomy: whereas in hunter-gatherer societies animals sacrifice themselves (Chapter 3), in pastoral societies the life and death decisions are determined by humans. To support his case, Ingold cites changes in material culture – the appearance of the whip, spur, harness and hobble – as manifestations of dominance, and draws upon Tapper (1988) and Tani's (1996) arguments that parallel animal domestication with slavery.

In broad terms there is much to commend Ingold's (1994) argument and it has remained central to debates about domestication ever since its publication. For instance Russell (2012) places much emphasis on the status of domestic animals as 'property' and how the concept of animal ownership allowed new hierarchical human–human relationships to develop based on ownership rather than sharing. However, others (e.g. Armstrong Oma 2010; Knight 2005, 4–6; Larrère and Larrère 2000) have up-ended Ingold's argument suggesting that, rather than representing a shift from trust to domination, domestication can be viewed in opposite terms, as a change from domination to trust. For while hunter-gatherer societies may trust that animals will sacrifice themselves, they do not necessarily have long-term or intimate relationships with them as individuals. Admittedly, the process of hunting is often very intense, and many modern hunting groups use mimesis, the mimicry of their quarry, to enable them to 'become' the animals they are hunting, so that they might seduce them into sacrificing themselves (Cartmill 1993; Willerslev 2007). That similar beliefs were held in the past has been argued cogently by Conneller (2004) based on the red deer antler frontlets from Mesolithic Star Carr in Yorkshire and wider evidence from Maglemosian rock art (see Chapter 3). However, far from a situation of trust, the distant interactions between hunters and their animals, together with the techniques of trickery used to capture them, arguably renders the relationship more callous, brutal and one-sided than is found in many pastoral societies.

Domestication brought humans and animals into very close, literally domestic, relationships; they shared houses, food and their day-to-day lives. Armstrong Oma (2010) believes that the relationships experienced under domestication were far more intimate, reciprocal and trust-based than Ingold (1994) gives credit. Importantly, Armstrong Oma makes the point that human–animal relationships are seldom so black and white as the trust–domination dichotomy implies. Instead, she invokes Larrère and Larrère's (2000) idea of 'social contracts' as a way to explain the more complex, multifaceted, multi-directional and situational character of inter-species relationships. This is not to romanticize the place of animals: Armstrong Oma accepts that, as with all relationships, power is important and unequally distributed, often with dire consequences for animals. Nevertheless, she maintains that no single, one-sided term can adequately account for the multiple context-dependent negotiations that occur during an animal's life and death.

Critically, authors such as Knight (2005, 4–6) and Armstrong Oma (2010) highlight that domestication need not have transformed ideas about ownership, trust and sharing in a negative sense (as is often portrayed). Instead, domestication brought animals and people together into new communities that experienced human–animal and human–human relationships that were increasingly intimate and just as likely to be founded on concepts of trust and sharing. From this perspective, domestication is not about objectifying animals but rather bringing them into communities with the resulting formation of closer mutual human–animal relationships. I do not necessarily believe that this was accompanied by a perception of animals as property but more as kin (as Russell suggested in 2007, and Midgley as early as 1978) and I can envisage a situation where many of the changes thought to characterize domestic animals – for instance size change and reduced flight response – could have been brought about as through the formulation of deeper and more meaningful relationships with animals.

To illustrate the above suggestion, I return once more to my chickens but also to Samuel the pheasant who, over the last 6 years, has episodically come to live in our garden where he hangs out with the flock. In truth, Samuel is, in all probability, many different pheasants – I suspect that 'Samuel' has been shot and eaten many times over. But, because all pheasants look pretty much the same, it is easy to believe that Samuel is one of our oldest animal acquaintances. Whilst my family know Samuel and wave to him as we travel round the UK, our relationship with him is certainly more distant than it is with the chickens – we know them intimately as individual characters, we know their behaviour traits and different personalities. However, not all of the relationships we have with our chickens were easily formed. Two years ago, we were fortunate enough to hatch five chicks: one yellow, one spotted and three black. It became very noticeable that we bonded with the yellow and spotted chicks, and gave them names, far more quickly than we did with the three black chicks, who we could not tell apart. When our neighbours got married, we decided that we would give them chickens as a wedding present; it was the un-named, more anonymous hens that we found easiest to give up.

As ever, it is important to consider how relevant my own experiences are to those of other societies in time and space. Anthropology offers some support for the suggestion that, in many agro-pastoral societies, animals are viewed as part of the community and that it is an individual's character and ability (or not) to form close, meaningful relationships with people that determines their life pathway. Amongst the Suri of Africa, for example, Abbink (2003: 342) has shown that cattle are a respected part of the human community and a precondition for all social relations: they are essential to the formation of individual identity, to family life, procreation and conflict resolution. Suri culture – in particular song, poetry and colour perception – is the product of human–cattle relationships. The Suri know the personalities and family histories of their cattle and most human personal names are shared with cattle; the names derived from their colour and markings (Abbink 2003: 347). However, as with our chickens, not all cattle are treated the same. It is the depth of relationship between the Suri and their cattle that determines their fate: pleasant and beautiful animals will be cherished but those that are aggressive or visually dull are more likely to be exchanged or sacrificed (Abbink 2003: 358). This is the kind of 'unconscious selection' that has been suggested by the likes of Zohary *et al.* (1998).

The possibility that coat colour and personality were equally important to the development of close human–animal relationships in the past would seem to be indicated by a number of different strands of evidence. For instance, early artistic representation of animals often illustrate coat colour/pattern in great detail, suggesting that the images are less depictions of a generic animal types (e.g. 'horse', 'cattle' or 'reindeer') and more portraits of specific individuals (Figure 2.3). Genetic studies equally point to the importance of animal appearance in past societies: although archaeological analyses of animal coat-colour are still in their infancy, examinations of horse coat colour (Ludwig *et al.* 2009; Pruvost *et al.*'s 2011) and pig pelage (Fang *et al.* 2009) all indicate a rapid and substantial increase in coat colour variation following shortly after the accepted dates of domestication.

Figure 2.3 Prehistoric African rock art – note that the cattle are depicted in far greater detail than the people.

Source: After Barker (2006).

Fang *et al.* (2009: 2) proposed that the motivations for the selection of coat colour may have been either a simple preference for 'exotic' appearance, a deliberate selection for reduced camouflage to facilitate animal husbandry or an attempt to distinguish domestic from wild animals. All of these explanations are plausible but a more social argument is that the development of coat colour variation was both instrumental in and the product of the intimate human–animal relationships that came about with domestication: essentially that people were more likely to have bonded with, and thus were more inclined to foster, those animals that were recognizable as individuals. In this way, the variety in appearance brought by domestication can be envisaged as the empowerment of *some* animals rather than the relegation of *all* animals to the status of slaves. I would propose that it was far later that animals became the subjects of humans, and later still that people actively stripped domestic animals of their individual identity through the creation of

Figure 2.4 Where's Wally? The individual becomes difficult to spot in a flock of similarity.
Source: Chris Graham, *Practical Poultry* magazine.

distinct breeds. Characterized by uniformity in conformation and appearance, a trait of modern factory-farmed animals, the similarity in appearance of 'breeds' must have served as a useful barrier to the formation of human–animal attachments (Figure 2.4).

This issue of appearance and character will be considered further in the section on the 'Agricultural Revolution' but, before that, I wish to consider the 'Secondary Products Revolution', which I feel can be reconsidered from the perspective of human–animal relationships.

Secondary Products Revolution

The concept of the 'Secondary Products Revolution' was the brainchild of Andrew Sherratt who, in 1981, published an article that inspired its own intellectual revolution. He proposed that domestication was initiated by people who sought to exploit animals solely for their 'primary products' (e.g. meat, fat, oil, skins, bones) accessible only once the animal was dead, and that it was not until much later that people recognized they could crop a wider range of useful resources or 'secondary products' (e.g. milk, traction/transport, wool) by managing animals in life. To support his argument, Sherratt provided a wide range of iconographic and artefactual evidence to demonstrate that material and visual culture for the exploitation of secondary products were absent until the fourth millennium BC in the Near East and third millennium BC in Europe, that is towards the very end of the Neolithic and into the Bronze Age. Sherratt (1981) utilized no zooarchaeological evidence in his original paper but this is understandable since large zooarchaeological datasets were scarce at the time. As new evidence from animal bone studies and scientific investigations has become available, zooarchaeologists (and Sherratt himself – 1983, 1997) have variously supported, reviewed, revised and countered Sherratt's core arguments (Anthony and Brown 2011; Bogucki 1984; Entwistle and Grant 1989; Greenfield 1988, 2010; Isaakidou 2006; Legge 2006; Marciniak 2011; Rowley-Conwy 2000; Vigne and Helmer 2007).

It is now generally accepted that the timing and circumstances surrounding secondary products exploitation were varied and complex, with no single or universal moment of adoption. Where detailed investigations have been undertaken most researchers have pushed the origins of secondary products utilization earlier into the Neolithic, with some scholars arguing that the term 'secondary' is misleading, especially where there is evidence to suggest that these products may have formed part of the initial process of domestication. For instance, in the case of domestic fowl it would seem that the principal motivation for their domestication and spread from Asia was never their primary products but rather those that could be cropped through life: their sound (recent genetic work on fowl from the Pacific islands suggested that sea-faring populations may have valued cockerels as 'fog horns' – Hannotte personal communication), perhaps for their eggs, probably for their feathers and certainly for cockfighting and divination (Tixier-Boichard *et al.* 2011; Serjeantson 2009; Zeuner 1963). In many cases the use of domestic fowl for their 'primary' products (meat) is a fairly recent innovation, as in Japan where there is little evidence that they were consumed regularly before the nineteenth century (MacDonald and Blend 2000: 496).

Dairying

Arguments about the utility of the terminology surrounding primary and secondary products has been made most forcibly in relation to cattle and sheep/goat dairying, which Vigne and Helmer (2007) believe occurred not only very early in the Neolithic but may actually have been the motivation for domestication. Their case is centred on detailed studies of caprine mortality profiles but it finds considerable support from chemical analyses of early Neolithic pottery sherds from central and northern Europe (Copley *et al.* 2003; Craig *et al.* 2005) and more recently from the Near East and southeastern Europe (Evershed *et al.* 2008): all have identified milk fat residues within ceramic fabrics, indicating that ruminant dairy products were being processed as early as the seventh millennium BC. However, as Russell (2012: 222) has pointed out, cattle and caprine domestication occurred in the *pre*-pottery Neolithic, so we are lacking evidence for the period that is most crucial to Vigne and Helmer's argument. To reinforce their case, Vigne and Helmer present a raft of interdisciplinary data that strongly suggest, but do not prove, milking occurred at the beginnings of domestication. Some of their examples come from outside the geographical and chronological focus of domestication, such as Balasse and Tresset's (2002) isotopic analyses of cattle teeth from the Middle Neolithic site in Paris-Bercy, France: here it was found that calves were maintained until just after weaning, suggesting that their presence was required in order to encourage the milk let down reflex of their mothers.

More convincing is Vigne and Helmer's application of palaeogenetic data in relation to the human development of lactose tolerance, which enabled people to digest unprocessed milk. They cite, amongst others, Burger *et al.*'s (2007) study that demonstrated beyond doubt that the adoption and spread of milking was responsible for the genetic adaptation for human lactase persistence, which became widespread during the Neolithic. This is an exciting example of human–animal co-evolution, again reinforcing the idea that the history of humans and animals cannot be separated (Holden and Mace 2003; Larson and Berger 2013).

The fact that people in the Mesolithic and Neolithic originally lacked the genetic mutation enabling them to drink large quantities of unprocessed milk (Burger *et al.* 2007) has been taken by some as evidence that widespread dairying could not have occurred until such point that lactase persistence had developed (Russell 2012: 227). However, when raw milk is processed into yoghurt or cheese the lactose is converted into a more digestible form and can be consumed without problem. Indeed, a more recent study (Leonardi *et al.* 2012) suggests that there are many examples where people who are genetically lactase *non*-persistent have been able to consume raw milk without any adverse effect. Perhaps most importantly, all juvenile humans are able to consume dairy products until they are weaned, at which point they undergo a physiological change that reduces ability to digest milk.

It has been suggested that children may have been one of the original motivations for dairying, the milk of tamed lactating animals being used to supplement the diet of children whose mothers had died or were unable to produce sufficient from their own breasts (Amoroso and Jewell 1963; Köhler-Rollefson and Rollefson 2002). Children might also have been inadvertently responsible for the genetic selection for lactase persistence. Ingram *et al.* (2009: 588) have suggested that the selection may have occurred during periods of famine, drought or outbreaks of diarrhoeal disease, when lactase persistent individuals would have benefited from their ability to take on extra fluid whereas milk consumption may have fatally exacerbated the conditions of non-persistent individuals. Such a scenario is consistent with the demographic change noted by Bocquet-Appel (2011a, 2011b) who cites childhood diseases, such as diarrhoea, as one of the causes of the high infant mortality witnessed in the Neolithic period. Such episodes of infant mortality could quite feasibly have resulted in positive selection for lactase persistence.

No single explanation is sufficient to explain global variation in lactase persistence/non-persistence but it would seem that, to some extent, the pattern reflects past differences in human–animal relationships and perhaps even climatically determined cuisine. For instance, it is interesting to note that lactase persistence is particularly common in the cooler northern hemisphere where milk would presumably have had a longer shelf-life than in warmer climes, so an adaptation to consume raw milk would have represented a significant selective advantage. In these temperate regions, milk appears to have been warmed and perhaps mixed with cereals as a kind of porridge (Dudd *et al.* 1999; Maier 1999) but, as heating does not make lactose more digestible, this may again have encouraged selection for lactase persistence. In the hotter environments of the

Mediterranean and Near East there would, perhaps, have been less demand for warm porridge, and milk if not drunk immediately, would have quickly turned sour. In these regions, therefore, conversion of milk into yoghurt and cheese was a logical form of storage but, as processing converts the lactose, this also negated the need for lactase persistence, hence frequencies of the necessary genetic mutation are low. It might be expected that the situation in Africa would be the same as that in the Near East; however there are regions of the continent that demonstrate very high levels of lactase-persistence. Leonardi *et al.* (2012) have suggested that these concentrations may reflect pastoral groups for whom access to sources of liquid in an otherwise arid landscape was a definite advantage.

Overall, I subscribe to Vigne and Helmer's (2007) view that milking was adopted early in the Neolithic. Yet none of their discussion, or anything that I have outlined above, really addresses the issues of why, or indeed how, people came up with the clever idea in the first place – as a student I puzzled over what would motivate a person to suddenly decide to milk an animal. However, once again, I suspect my confusion says more about the 'Western' society in which I live than that of the Neolithic. In the modern 'developed' world the milking of animals usually takes place behind closed doors, just as human breastfeeding in public is often considered embarrassing or, worse, illegal (Raju 2006). It takes only the briefest review of the anthropological literature to realize how weird Western culture is by global standards, and presumably in terms of global history. Although studies are few, there is considerable evidence for human–animal co-milking, whereby women breast-feed animals (Simoons and Baldwin 1982) and animals suckle children (Knight 2005: 10–11). The Bishnoi community in Rajasthan is again a good example of this kind of human–animal bonhomie, as shown by the scene in Figure 2.5, recently captured by photographer Himanshu Vyas, of a woman who simultaneously breast-feeds a wild deer and her daughter (Kothari and Sharma 2013). All who have considered the subject (Köhler-Rollefson and Rollefson 2002; Milliet 2002; Simoons and Baldwin 1982) have implicated this kind of reciprocal milking in the origins of domestication. Certainly such a scenario would be consistent with my earlier suggestion (see pages 34–5) that increasingly intimate human–animal relationships, rather than simply the desire to exploit animals for meat or milk, were both the product of and motivation for domestication (Figure 2.5).

As with the issue of domestication, the focus on elucidating the origins of milking has, to some extent, diverted attention from the social implications of dairying. Admittedly, anthropological studies of milk production are thin on the ground but there is sufficient evidence to show that, in pre-industrial societies, the production, distribution and consumption of dairy products are highly important components of daily life (e.g. see the many wonderful papers in Lysaght 1994, and also Crate 2008). Dairying is often central to community cohesion, social differentiation, identity formation, and it can also serve to create artistic and material culture, evidenced by the considerable folk-lore, folk-music, art and artefacts associated with dairying (Abdalla 1994; Crate 2008; Ling 1997; Salomonsson 1994). In addition, gender roles are frequently defined through dairying: although there are no universal standards, cross-cultural analyses suggest

Figure 2.5 A moment of human–animal intimacy.
Source: Himanshu Vyas, redrawn by Sykes.

that, in general, women are more closely aligned with milking and milk processing (Crate 2008). This is especially the case at the domestic level, where dairy products can become correlated with a woman's personal esteem and social standing (Skjelbred 1994). Archaeological, iconographic and historical evidence indicate that the association between women and dairying has considerable time-depth. For instance, Davidson (1998: 30–5) has argued convincingly that connections between milk animals and women (particularly female deities and saints) can be traced consistently from third-millennium Mesopotamia all the way through to post-medieval Ireland. Whilst this association is confirmed by other scholars (see papers in Lysaght 1994), it is clear that men take a more prominent role where dairy production is on a more industrial level (Yentsch 1991: 134).

All of these interesting human dynamics – the very things that most archaeologists joined the profession to try and understand – are largely ignored in the bone-focused zooarchaeological literature. Once again, this highlights the need to take a broader approach if we are to understand the human–animal relationships reflected in zooarchaeological assemblages.

Riding and Traction

The horse is another example where all available evidence suggests that domestication and the use of individuals for 'secondary' products (milking but in this case also riding) were closely linked. Outram *et al.*'s (2009) study of the horses from the Botai culture in Kazakhstan, long acknowledged as the centre for horse domestication, provides three compelling lines of data to indicate that domestication, milking and riding occurred contiguously in this region during the mid-fourth millennium BC. First, osteometric analysis demonstrated the Botai horses to be akin to bones from domesticated Bronze Age individuals rather than those from wild horses of the Palaeolithic. Second, lipid analyses of ceramics proved conclusively that both horse meat and milk were being processed in Botai culture pottery. Third, Botai horse premolars were found to exhibit parallel-sided bit wear (Figure 2.6), which experimental studies have shown is a trait only of bridled animals (Bendrey 2007).

In this case, the use of bridles or harnesses would seem to support Ingold's (1994) belief that domestication equated to human domination, with animals being brought quickly into slavery by people. However, Armstrong Oma (2010: 180) draws upon the work of Game (2001) to highlight that, for riding to 'work', the horse and human must trust each other, acting in unison and responding to each other's movements as if a single being. This must have been particularly true for the period before the invention of saddles and stirrups, when there would have been little physical separation between horses and their riders – the two must have been very much attuned to each other. For this reason, I find the parallel-sided bit wear one of the most eloquent and, importantly, unambiguous expressions of the

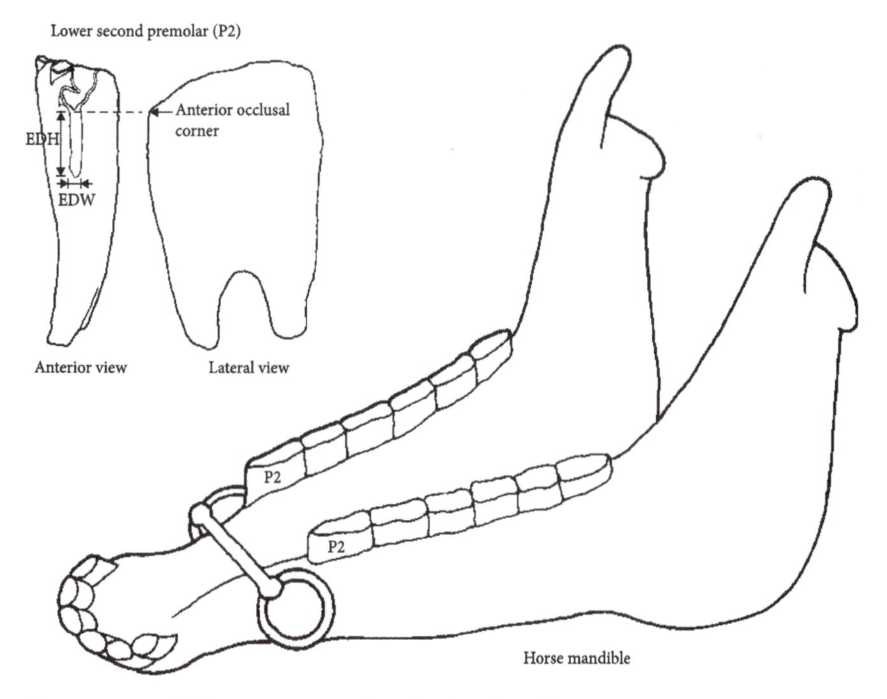

Figure 2.6 Parallel bit wear – a trait found only in bitted horses.
Source: Robin Bendrey.

new intimate human–animal relationships that emerged during the Neolithic period.

Not all new human–animal relationships are so easy to detect, the use of cattle for traction being a case in point. The presence of plough cattle is often claimed where assemblages contain an abundance of adult animals, the assumption being that they were maintained for reasons other than meat, most probably secondary products such as traction. At first glance the argument seems persuasive, especially when the age data are supported by a high incidence of palaeopathological traits that might feasibly be traction-induced: e.g. extra bone growth (exostosis), lipping, eburnation and/or grooving on the proximal or distal articulations of metapodia, phalanges or weight-bearing joints (De Cupere *et al.* 2000; Isaakidou 2006: 97). However, these pathologies, although aggravated by heavy work, can occur independently. In fact, many are age-related, so a higher frequency would be expected in an assemblage dominated by adult cattle – it need not indicate their use as plough animals. This problem of equifinality means that, currently, convincing arguments for cattle traction require the integration of multiple strands of evidence considered over the long term. Johannsen's (2006) study of cattle in Prehistoric Scandinavia is exemplary in this regard. Beginning with an analysis of pathological traits in the foot bones of wild aurochsen, Johannsen then makes detailed comparisons with assemblages of domestic cattle from different periods of the Neolithic. Using this technique he was able to demonstrate that whilst the Early Neolithic cattle showed similar patterns of pathology to the aurochsen assemblages, a far higher incidence of pathology was noted on the Middle Neolithic cattle. Importantly, this rise in stress-related traits was coincident with a clearing of the landscape, evidenced by the pollen record, the appearance of new and larger settlements, and the first appearance of the ard/scratch plough. These independent lines of enquiry all suggest that, in this region, cattle became important providers of traction during the Middle Neolithic, in the second half of the fourth millennium BC, a time of considerable social upheaval and change. Johannsen (2006: 44) draws comparisons between the Scandinavian data and that for Switzerland, where similar trends are observed in the artefactual, landscape and zooarchaeological records of the middle of the fourth millennium. In northern Europe, at least, it would therefore seem that traction was a secondary product, chronologically speaking, with cattle being utilized for meat and milk at an earlier date.

Of all the new human–animal relationships of the Neolithic period, the use of cattle for ploughing is, I feel, most consistent with Ingold's (1994) argument for human domination of animals. For although Armstrong Oma (2010: 181) states that ploughing requires animals and handlers to have mutual rhythm in their movement, and that it makes economic sense to employ trust and benevolence towards animals rather than brutality, the history, anthropology and technology of yokes and ploughing equipment (Conroy 2004) provide little reassurance about the welfare afforded to Neolithic traction animals. Where an increase in animal palaeopathology is demonstrably linked to ploughing, the conclusion must be that the individuals were being encouraged to behave in a way that caused them discomfort.

That said, we must query whether the discomfort felt by animals was any more than the levels experienced by contemporary humans: multi-period studies of osteoarthritis and musculoskeletal stress markers (MSM) in humans have often demonstrated an increased prevalence, particularly in female and adolescent skeletons, following the introduction of agriculture (Eshed *et al.* 2004; Papathanasiou 2005; Ponce 2010). This suggests that everyone – humans and animals – started to work harder in order to produce and process new foods to sustain the whole (human and animal) community. However, the trend towards increased stress markers in Neolithic human skeletons is far from universal; other studies have found that agriculture brought a decline in human MSM (e.g. Eshed *et al.* 2004). It is, therefore, necessary to take a local perspective on the issue and perhaps greater attention should be given to the analysis of relative rates of pathology in humans and animals to see if it is possible to identify situations where humans started to benefit by offloading burden onto animals – this would be an interesting area for future research.

Wool

Wool is the one animal product that we can be fairly confident was exploited secondarily to domestication; the coats of wild sheep consist more of hair and kemp than wool, and it would have taken generations of selective breeding to produce the woolly sheep familiar to us today. Osteological sheep remains are a poor medium for understanding the evolution of wool production, although it is widely assumed that assemblages composed of adult animals are indicative of wool economies, the idea being that these animals were kept for several years in order to gain the maximum number of wool clips (Bökönyi 1977: 25). As was seen in Chapter 1, age profiles can reflect a wide variety of factors and the economics of wool production need not be the reason for maintaining sheep to old age. Better sources of evidence are the paraphernalia associated with spinning and weaving, although it must be remembered that, in many areas of the world, textile production based on vegetable fibres considerably pre-dated the creation of wool-based fabrics. In the Near East spindle whorls and loom weights are found increasingly from 5500 BC onwards (Wild 2003a: 43) but the presence of coarse-coated woolly sheep is not attested until around the mid-fourth millennium BC, indicated by the fibres in textiles recovered from Shahr-i Shokhta in eastern Iran and Novosvobodnaya in the North Caucasus (Gleba 2012: 3644). Actual wool remains are, therefore, the most secure line of evidence but they are rare in the archaeological record, their representation highly dependent on favourable preservation conditions.

A number of studies examining archaeological wool were conducted by Ryder (e.g. 1969, 1983, 2005) who argued that the coats of early domestic sheep were similar to those of their wild counterparts, with an outer coat of long kemp fibres and a finer woolly undercoat of shorter fibres, that people would have obtained by plucking rather than shearing (Ryder 1983: 45–7, 95). Gradually, sheep were bred for more uniform fleeces, with coats of medium fibres becoming predominant in Iron Age textiles dating to around 1500 BC. Ryder (1974) suggested that fine fleeces appear around the Mediterranean in the Classical period, becoming more widespread geographically during the Roman period (Ryder and Gabra-Sanders 1985: 130). Gleba's (2012) detailed analysis of archaeological textiles from Apennine Italy has largely confirmed Ryder's overarching evolution of wool types; however, it has also provided a more nuanced chronology for the region, suggesting that uniform, but coarse, woolly coats began to emerge around the Late Bronze Age to Early Iron Age transition.

Understanding when sheep first began to be used for wool is exceptionally important and many zooarchaeologists have examined the origins and spread of wool production (for a review see Greenfield 2010). However, most analyses tend towards the sterile with attention focusing on the detection of patterns in the zooarchaeological data that might feasibly indicate wool production rather than considering the social ramifications that they represent. Schneider's (1987) *The Anthropology of Cloth* demonstrates just how significant an oversight this is; her article is a master-class, highlighting the role of textile production in the consolidation of social relations, the negotiation and expression of identities and ideologies, as well as the mobilization of power.

To some extent zooarchaeologists working in the Near East can be forgiven for overlooking the social significance of wool, since textile manufacture was already well established before woolly sheep came on the scene and it is difficult to say how much impact the switch from weaving plant to wool fibres would have had (Good 2012: 337; Richmond 2006: 204). Nevertheless, the issue still deserves consideration and McCorriston (1997) has argued that, in Mesopotamia at least, the transition was accompanied by increased social complexity, enabled by the advantages of time, productivity and versatility that wool offered over flax. By comparison to flax, which requires weeks of retting and drying, wool is quick to process, being ready for use within as little as a few hours. Wool is also naturally more varied in colour than flax and far easier to dye than plant fibres, allowing for greater creativity in textile design (Algaze 2009: 80).

Across much of northern Europe, the arrival of woolly sheep was probably a more significant phenomenon, especially since there is little evidence for woven fabrics prior to their introduction. Spindle whorls and loom weights are rare in northern Europe during the Neolithic period. It is possible that examples made of organics have not survived archaeologically but it is notable that ceramic varieties become more common on settlements at approximately the same time that woollen textiles also start to appear (Barber 1991: 4; Edwards 2006: 7). In Europe the earliest example – a textile woven from mixed plant and wool fibres – comes from Wiepenkathen, near the mouth of the Elber, and dates to *c.* 2400 BC (Jørgensen 2003: 55–7). According to Jørgensen (2003) weaving paraphernalia becomes more abundant after this date.

Finds of wool are very rare for Prehistoric Britain, most probably due to the issue of preservation. However, spinning and weaving tools occur regularly from the Late Bronze Age onwards which, together with the dramatic increase in the archaeological representation of caprine bones, suggests woolly sheep were introduced around this period (Bradley 1978: 46; Serjeantson 2011: 96). If this was the case – that animal husbandry shifted from cattle to an emphasis on sheep, and textile production became more common – we may assume that this was accompanied by considerable changes in human–animal relationships, daily practice and experience. At the most basic level, the advent of new textiles that were naturally multi-coloured but also had the capacity to retain artificial dyes (which Brandt *et al.* 2011: 212, argue were employed from the very beginnings of woollen textile production) would fit with the wider evidence for the explosion of colour identified for Bronze Age Britain (Jones 2002). However there is good reason to suppose that it was not simply the end product but the whole process of production, distribution and consumption that would have rendered woollen goods important mechanisms for the negotiation of social relations and gender roles.

Some scholars suggest that gender roles became increasingly defined during the Bronze Age and into the Iron Age (Robb 1997; Treherne 1995; Whitehouse 2012: 497), and textile production would certainly have helped with this process. Schneider (1987) has shown that, today, across cultures, spinning and weaving are predominantly, though certainly not exclusively (Costin 2012: 188), the domain of women. There is clear evidence to suggest that the same was true in the past (Barber 1995; Costin 2012). From the very first examples, iconographic representations of textile production overwhelmingly depict women (Barber 1991, 1995). For instance, the hairstyles of all the individuals shown spinning and weaving on the fourth-millennium cylinder seals from Mesopotamia indicate that they are women (Barber 1991: 57). The same is true for the spinners shown in a third-millennium mosaic from Mari, Syria, and in the first-millennium reliefs from Susa in Iran and Maras in Anatolia (Barber 1991: 58–9). For northern Europe the first representation dates to the Iron Age: the seventh-century BC Hallstatt vase recovered from Sopron (western Hungary) shows one woman spinning and another weaving (Barber 1991: 55). Similar scenes occur on Greek vases dated to the sixth century BC (Barber 1995: 82). The theme had not changed by the Roman period, with tombstones depicting widows associated with balls of yarn and spindles, reflecting belief that spinning was a sign of a dutiful wife (Wild 2003b: 82).

The historical evidence shows a similar female bias, with women associated with spinning and weaving in Old Assyrian (Veenhof 1972), Babylonian (Dalley 1980) and Linear B texts (Wild 2003a: 47). Within Greek mythology weaving is associated with women such as Calypso, Circe, Arache, Minerva, the last being the patroness of weaving; and the three Fates (all of whom were women) spun thread to be the weft woven in and out of the warp that forms a person's life events (Kehoe 2000: 136). Similarly, the Valkyries of Viking Age Scandinavia strung their looms with human intestines and the weave of the cloth controlled battles and decided the fate of warriors (Davidson 1998: 117–18). Within the Sagas, textile production is an important theme, often appearing symbolically alongside scenes involving violence, death and battle (Enright 1996). This is also reflected in the weaving paraphernalia from medieval Scandinavia, with the presence of weaving *swords* and *spears*. The parallel between violence and weaving is not unique to Scandinavia; for instance, excavations of post-classic Aztec sites have yielded spindlewhorls decorated with the same design used on shields (Kehoe 2000: 135).

On the basis of grave-goods evidence from Iron Age Italy, Robb (1997: 51) has argued that spinning/weaving were, for women, the equivalents of warrior status for men. This concurs with Gilchrist's (1999: 51) suggestion that, in Anglo-Saxon and Medieval England, women conveyed their femininity through the process and artefacts of spinning and weaving. Certainly in these periods, and across cultures, spindle whorls and other weaving paraphernalia are typically found in female graves (Barber 1995: 288–9). The pattern is consistent from third-millennium BC sites in Turkey and northern Iran (Keith 1998: 499), through Iron Age and Early Roman Italy (Gleba 2009; Holloway 1994: 115; Sestieri 1992), to Anglo-Saxon England (Gilchrist 1999: 50) and Viking Age Denmark (Jesch 1991: 14).

Whilst overarching trends are observable in the archaeological, historical and iconographic record, this is not to suggest that the meaning attached to the activities of textile production was cross-culturally uniform. Nor should it be assumed that gender relations were formed according to static male–female oppositions in daily practice: e.g. women as weavers, men as warriors. Schneider's (1987) work makes clear just how dynamic the relationship between men, women and textiles can be, with fabrics involved in the most important rites of life and death for both men and women, young and old: cloth is used in the rituals of birth (e.g. swaddling newborns), marriage (e.g. as bride veils, for uniting bride and groom, enclosing the wedding bed) and death (enshrouding the cadaver).

In many societies, both past and present, woollen textiles are prepared by women in anticipation of marriage (Schneider 1987: 410) with fabrics frequently bequeathed by older women to younger women of marriageable age, an act that serves to unite the inter-generational female line. According to Gilchrist (2012: 127) this pattern of behaviour is recorded in medieval English wills, which contain examples of women bequeathing their own bedding to young women 'against the marriage'. This would seem to confirm Gilchrist's suggestion that textiles, in particular bedding, formed part of a bride's dowry, which is perhaps unsurprising given that the bed is central to marriage and child-birth (Gilchrist 2012). Similarly Dalley's (1980: 55, 63–4, 66–74) study of Old Babylonian dowries has highlighted that wool, weaving tools, textiles and even shrouds were frequent components of dowries, again indicating the social significance of woollen fabrics.

Having outlined just a small fraction of the literature pertaining to wool and textile production, I no longer feel that my ability as a zooarchaeologist to analyse collections of sheep bones and conclude that 'sheep were managed for wool' (Sykes 1998–2012, any report) does justice to the assemblages that I study. Wool has the capacity to, quite literally, weave societies together but to understand how, or whether, this occurred in the past requires that animal bone data are themselves woven into a much greater tapestry of cultural evidence.

Agricultural Revolution

Unlike the Prehistoric period, where animals and zooarchaeological studies have been given a privileged position within archaeological discourse, markedly less attention has been given to animals in relation to the historic period. Fairly recently, however, European zooarchaeologists have managed to capitalize on the buzz word 'revolution', applied by economic historians to the agricultural intensification and stock improvement of the early modern period. The Agricultural Revolution was originally viewed as the country-bumpkin cousin of the Industrial Revolution, both occurring between about 1760 and 1840 but, as is the case with all revolutions, the exact dating and form of the Agricultural Revolution have been the subject of debate with scholars seeking to find ever earlier examples of its first inception.

Historians such as Kerridge (1967) and Beckett (1990) argued that shifts in agriculture and animal husbandry associated with the revolution had a far deeper history, their origins dating back to at least the sixteenth century with transformations continuing until the end of the nineteenth century. By the mid-1990s a number of zooarchaeologists, notably Simon Davis and Umberto Albarella, had begun to engage heavily with historical debates about the Agricultural Revolution, collaborating with historians (e.g. Davis and Beckett 1999) to review the evidence from a zooarchaeological perspective (see also Albarella 1997b; Davis 1997). By synthesizing animal bone data from a wide variety of sites they were able to highlight dramatic shifts in the size and age structure of the main domesticates. Cattle, sheep, pig, domestic fowl and horse were all shown to demonstrate substantial increases in stature, suggesting that considerable and widespread investment was being made to improve breeding stock, most probably in the hope of achieving greater carcass weights. An emphasis on meat production was certainly indicated by ageing data for cattle, pigs and geese, which showed a contemporary increase in the exploitation of juveniles (Albarella 1997b). By contrast, age profiles for sheep and horse indicated a shift towards older ages, the animals presumably being kept for wool and traction (Albarella 1997b). All of these changes, and more besides, strongly suggest the kind of economic transformation that might have accompanied the Agricultural Revolution; however, they appear to have occurred significantly earlier than traditionally accepted, with many of the changes appearing in assemblages dated to the fifteenth century. More recently, other zooarchaeologists (e.g. Thomas 2005a, 2005c; Thomas *et al.* 2013) have documented clear changes in animal size and age structure apparent at sites, such as Dudley Castle in the West Midlands, dated as early as the fourteenth century. Although Thomas (2005a, 2005c; Thomas *et al.* 2013) does not explicitly tie these shifts to the Agricultural Revolution, they are implicated as an early manifestation of the economic changes that later spread across the country.

Zooarchaeological investigations of the Agricultural Revolution, or at least medieval/post-medieval animal improvement, are now found across Europe from Scandinavia and Iceland (Hambrecht 2006, 2009; Puputti 2008), France (Clavel *et al.* 1997) and Iberia (Davis 2008b). In each region the timings for the emergence of new animal husbandry regimes vary – for instance in Portugal a pre-fifteenth century increase in cattle is noted with sheep showing a rise in stature even earlier, whereas in Scandinavia increases in size do not occur until the seventeenth and eighteenth centuries. Regardless of their timings, explanations for the appearance of new husbandry regimes are consistently linked to wider changes in diet and economy; for instance they are frequently interpreted as reflecting a new emphasis on productivity in a period when high demands for meat were generated by growing urban populations (Albarella 1997b; Davis and Beckett 1999; Hambrecht 2006; Puputti 2008). Often interpretations are skilfully conjured from the integration of historical, landscape and human demographic evidence to show that new animal regimes were most probably born from periods of population crises and social change, such as those brought by the Black Death (Albarella 1997b; Davis 2008b; Thomas 2005a, 2005b). I find these explanations entirely convincing and do not wish to critique them; however, I feel there is potential for examining these transformations in terms of human–animal relationships and considering what they might reveal about wider cultural ideology.

Within the social sciences considerable attention has been given to the social significance of animal husbandry in the late eighteenth and nineteenth centuries, when the development of prize breeds of livestock became a pursuit of the aristocracy. These gentlemen farmers were keen to demonstrate their philanthropy towards yeomen of more modest means, who lacked the resources to experiment with agricultural technology, and towards the national population for whom meat was becoming a valuable commodity (Ritvo 1987: 48). However it is clear from textual evidence and visual culture, in particular the rise of livestock portraiture, that motivation for intensive animal husbandry went far beyond issues of economics and diet (Ritvo 1987: 58; Quinn 1993). Without doubt, livestock became the embodiment of their owners' status with animal markets and shows serving to display not just animals but also the achievements of their masters. But the importance of livestock breeding went further still: Ritvo (1987) has made the case that pedigree animals, with their documented ancestry and racial purity, represented the very ideals that members of the elite wished to emphasize about themselves. Anxiety about the dilution of the bloodline combined with a desire to differentiate between emerging breeds, encouraged agriculturalists to establish breed standards that were laid out in both texts and artistic representation – these formed the models to which good husbandsmen sought to adhere without deviation (Quinn 1993). It is my belief that these breed standards and the drive to produce identical animals also helped to erase any perception of animals as individuals, instead seeing them lumped together as 'breeds' that were identifiable as the efforts of a particular human individual. The fact that different, visually separable, breeds were considered suitable for one product or another (milk, wool, meat, eggs, fighting) can be seen as an early form of product branding – at this point animals were truly commercial products. Of course, some animals, such as the famous Durham Ox, achieved great individual status as national icons; however such examples are remarkable by their rarity.

Given the cultural significance attached to intensive breeding in the early modern period, it seems appropriate to consider whether earlier evidence for animal improvement, as yet detected only in the zooarchaeological record, was also accompanied by equally significant but undocumented changes in attitudes to animals. It seems, for instance, no coincidence that the earliest zooarchaeological evidence for animal improvement in England dates to the fourteenth century, the very moment that the gentry (a word that is itself related to the term 'gene' and means, essentially, 'good breeding') were beginning to emerge as the new social class (Coss 2003; Nicholson 2011). The gentry occupied the position below the nobility, to which they aspired, and above the yeomen and merchants, from which they sought to differentiate themselves. As Nicholson (2011) has pointed out, their place in English social hierarchy was ambivalent, demarcated from the lower classes only by the possession of an education, a coat of arms and, importantly in this context, a claim to lineage. Whilst human bloodlines take time to create, a respectable ancestry can be generated more quickly via animals which may yield many generations in the same time period required for human procreation. Tied explicitly to the landscape (Quinn 1993: 152) particularly in the case of regional breeds and types, animals can act as a nexus for validating claims to land and power; a useful asset in a period when the gentry were seeking to do both.

An alternative possibility, but one also tied to the gentry, is that the emergence of an ambitious social group who were capable of entering the nobility through marriage or merit, might have caused anxiety amongst their immediate social superiors and it may have been the nobility who took the first steps towards creating animal expressions of their ancestry. With this in mind it is noteworthy that some of the earliest evidence for animal improvement comes from an elite site, Dudley Castle in the West Midlands. Furthermore, it coincides with a moment that the castle passed into new hands after the previous owner had died without an heir (Thomas 2005c: 20). Ideas of primogeniture and the importance of fertility must surely have been a keen concern of Dudley Castle's new owners and what better way to demonstrate procreative force than to establish a substantial herd, both in terms of stature and size – something that they clearly did. Perhaps similar concerns, not just diet and economics, were the impetus for widespread medieval (and perhaps also Roman) animal improvement.

Summary

The animal revolutions considered in this chapter are all rather elusive and I suspect that very fact is what makes them such attractive foci for research and a convenient peg on which to hang successful funding applications. However, I tend to agree with the sentiment of Halstead (1996: 306) when he suggested that zooarchaeologists are probably more concerned with the issue of domestication (but I would also add the origins of secondary products and improvement) than anyone who was actually involved in the phenomenon. Similar to Whittle's (2007: 623) argument that it is, perhaps, time to abandon the labels of 'Mesolithic' and 'Neolithic' as there are simply too many different versions of both, I wonder if we should take a break from our pursuit of other revolutionary labels.

I am not suggesting that we should stop investigating the periods in which the 'revolutions' occurred, far from it – I agree that from as early as the twelfth millennium BC, human society began to transform with new ways of relating to animals, which must surely have created and reflected a new worldview. However, I feel that much effort has been expended trying to divide and define what is a tangle of great complexity and variety that cannot be shoe-horned into neat, opposing boxes of wild/domestic, hunter-gatherer/farmer, nature/culture, trust/domination, primary/secondary, primitive/improved.

To be fair, these opposing labels are important. This is because they are, essentially, the 'folk taxonomy' of the archaeologist (Chapter 1, page 6) – they appeal to and reflect our worldview, if not the societies that we study, and that is perhaps the best reason for maintaining them. But there is a need to move away from the study of dichotomies towards the investigation of more nuanced and, I think, exciting issues of whether particular animals were avoided, observed, hunted, tamed, abused, herded, milked, ridden, selectively bred, maintained as pets – the list goes on and can be reconfigured and hybridized in any number of different combinations. All these forms of interaction represent dramatically different human–animal–plant–landscape relationships with equally varying implications for human experience, practice, and thus the very creation of culture and worldview. However, as with all labels, those listed above are just that: labels. The same actions (e.g. animal breeding or taming) can be carried out for very different reasons and with different context-specific meanings: hunting a red deer with a bow and arrow in the Mesolithic period is unlikely to have been experienced in the same way as an identical act undertaken in the later medieval period. By way of demonstration, the next chapter will examine this very issue: how hunting – its culture, landscapes and meaning – varied through time.

Chapter 3

WILD ANIMALS AND HUMAN SOCIETIES

Some of the most pivotal moments in human history have been characterized by shifts in the relationship between people and wild animals. As was outlined in the last chapter, domestication not only gave rise to the category 'wild' but, with the arrival of new domestic animals, it also meant that the killing of wild animals was no longer necessary. Yet people continued to do so: in some periods wild animal killing was a rare occurrence, in others it became a frequent pursuit, evidenced by the art and literature of the time (e.g. Almond 2003; Anderson 1985; Barringer 2001; Cartmill 1993; Cummins 1988; Griffin 2007; Hamilakis 2003; Marvin 2006).

Today, the killing of wild animals is often a highly emotive issue that divides opinion along cultural, social, political and ideological lines, encouraging people to campaign vehemently according to their individual beliefs and identities (e.g. Marvin 2000). Our attitudes towards wild animals say a lot about us, and it is likely that the same was true in the past. Despite this, archaeologists have traditionally given little consideration to how wild animals were perceived and engaged with by the people who lived alongside them. Hunter-gatherer societies are widely considered as 'optimal foragers' who killed the largest available animals to obtain the maximum amount of meat (Mithen 1990: 44). It is assumed that, once people had access to domestic livestock and derived most of their protein from them, wild animals were disregarded and utilized only in times of need as a 'risk buffering' strategy (O'Shea 1989; Grant 1981). It is true that wild foods represent a dietary resource that agricultural communities turn to in times of famine; however, farming societies are seldom ambivalent to wild animals and even where they are not exploited, they always carry social meaning (Cartmill 1993; Hamilakis 2003; Kessinger 1989; Morris 1998).

In essence, human responses to wild animals reflect individual, social and cultural attitudes to the natural world; they map how humans perceive their place in the cosmological order. In this chapter, by way of a case study, I will review the English evidence for human–wild animal relationships from the Mesolithic to the medieval period to examine how they might reveal shifts in cultural worldview. Before we consider the evidence, however, we need to define the terminology.

Definitions of Hunting and the Wilderness

All too often we use the word 'hunting' to describe the act of killing wild animals but, in fact, very few acts of animal-killing can actually be classified as hunting. According to Cartmill's now classic volume (1993: 197, my emphasis):

> Hunting in the *modern* world cannot be understood as a practical means of obtaining cheap protein. It is symbolic behaviour and . . . can be understood only in symbolic terms . . . hunting is not just a matter of going out and killing any old animal; in fact very little animal-killing qualifies as hunting. It must be a *special* sort of animal, killed in a *specific* way for a *particular* reason.

Indeed, Cartmill goes further to suggest that animal-killing can be classified as hunting only if it meets a number of specific criteria:

1. that the animal is wild, not tame;
2. that the animal is free;
3. that the kill is violent;
4. that the violence is inflicted directly, it cannot be mediated by a trap;
5. that the kill is premeditated.

Whilst Cartmill's work is both European-centric and focuses more on farming than hunter-gatherer communities, his criteria are useful in general terms. The definition is flexible enough to include the killing of tame or even domestic animals that cross the boundary and become conceptually 'wild' through the activity of hunting – for instance, at Neolithic Durrington Walls (UK) several domestic cattle and pig bones had flint arrow/spear tips embedded in them, leading Albarella and Serjeantson (2002: 44) to suggest that these domesticates were ritually hunted. Furthermore, although we tend to use the term 'hunting' in relation to the capture of land mammals, Cartmill's (1993) criteria may be equally applied to some activities associated with fowling or fishing (Mylona 2008).

Given the emphasis on direct violence, and bearing in mind that hunting tends to employ the same tools used in warfare (bows and arrows, swords, knives and spears), it is perhaps unsurprising that hunting is, across cultures, frequently linked to martial ability and, in particular, masculine identity. As a highly visual and, usually, ritualized performance, hunting is also often used by social elites as a mechanism of social control but also of power legitimization (Barringer 2001; MacKenzie 1988: 55–88). Within farming societies, the power linked to hunting is often derived from wider perceptions of landscape (Chapter 5) and in cultures where the concept of land ownership exists, wild animals are often deemed to be the land-owner's 'property', with social or legal restrictions being placed upon their capture and consumption (Morris 1998: 93–7; Sykes 2006a). Another reason for the significance of wild animals is that many agricultural societies view the landscape beyond their homes and fields as a different realm, perhaps best described as the 'wilderness'.

The term wilderness is used here not to suggest some primordial landscape untouched by humans; instead the word is defined according to its original meaning: etymologically it comes from the Old English *wildeoren*, meaning '*wildern*' (wild or savage) and '*deor*' (animal), denoting an area where wild animals are found in greater numbers. As Hamilakis (2003: 240) has argued, the social rhythms of the everyday do not apply within the wilderness and to venture into this landscape is therefore to move within a less familiar geography. Such landscapes are frequently perceived to be the lairs of predatory animals and supernatural beings, dangerous or sacred places where past and present and life and death may be merged (Cartmill 1993; Ingold 2000: 84; Morris 1998: 104; Semple 2010). In her cross-cultural analysis of traditional societies, Helms (1993: 153–7, 211) has shown that it is the hunters' ability to travel between the boundaries of these different worlds, bringing back animals and other things from the outer realm, that is often regarded as a sign of power and those who do so are frequently conferred with a shamanic status or supernatural authority.

Given the anthropological evidence for the social importance of hunting, it seems strange that archaeological studies tend to consider wild animals only in terms of food procurement strategies. Widespread application of the optimal foraging model within an archaeological context has generated many energy-efficiency conclusions; however, biological models fail to account for the socially and culturally dictated behaviour of humans, and it is these factors that will be the focus of this chapter.

Sources of Evidence for Understanding Human–Wild Animal Relationships

Any attempt to consider how relationships between humans and animals changed over such a long period must first outline the sources of available data: in this case they are highly uneven due to temporal variations in landscape and environment, faunal populations and cultural practice, as well as variations in archaeological preservation and recovery methods.

Perhaps unsurprisingly evidence for the Mesolithic period is most scarce (Schulting 2013: 313). In England zooarchaeological material comes from a small range of sites, such as Star Carr in Yorkshire and Thatcham in Berkshire, so it is necessary to draw comparisons with contemporary evidence from elsewhere in Europe, as well as from anthropological discussions. The same is true for the Neolithic and Bronze Age, for which archaeological data are more forthcoming but the representation continues to be patchy and derived mostly from ceremonial contexts – enclosures and funerary structures – rather than domestic settlements (Schulting 2013: 317; Serjeantson 2011).

Human engagement with wild animals can, in some cases, be gleaned through isotopic analysis of human remains, carbon and nitrogen analyses being particularly useful for highlighting whether or not people were actively procuring and consuming marine resources. Such analyses are, however, dependent on mortuary practices: human remains may not always be available for scientific study, particularly in the case of cultures that employed excarnation or cremation rather than inhumation, as is found in some geographical regions in the Neolithic, Bronze Age, Iron Age, Roman and Early Anglo-Saxon periods (Bond and Worley 2006; Madgwick 2008; Redfern 2008; Smith 2006). Funerary traditions are, in their own right, important sources of information about attitudes to wild animals and the natural world (e.g. Fowler 2011) and so funerary rites will also be considered in this chapter.

The value of funerary contexts may be further increased where grave goods are incorporated into human burials, especially where these goods include weapons or the remains of wild animals. Grave goods are traits of (some) Bronze Age, Iron Age, Roman and Early Anglo-Saxon mortuary rituals but the restoration of Christianity in the middle of the seventh century AD saw a change in mortuary traditions, with a decline in grave goods from the eighth century (Härke 1989). The eighth century would also appear to be the cut-off point for most archaeological scientists, who appear to have less interest in undertaking isotope analyses of the more 'modern' human remains: there are many fewer studies of historic-period populations than those focusing on the Prehistoric period (Müldner and Richards 2006: 228). Nevertheless this reduction in scientific data is countered by a greater availability of documentary and place-name evidence, which become more common from the middle Anglo-Saxon period onwards.

Animal remains are the one source of data that are available for all of the periods under consideration but their analysis is not without problem. Recovery of material during excavation introduces considerable issues for investigations of wild animals, particularly birds and fish whose bones will be recovered reliably only with sieving, a practice that excavators do not always adopt as standard. Identification may be equally problematic, especially between closely related species such as wolves (*Canis lupus*) and dogs (*Canis familiaris*), wild boar (*Sus scrofa*) and domestic pigs (*Sus domesticus*), and pheasants (*Phasianus colchicus*) and domestic fowl (*Gallus gallus*). Even where specimens can be identified to species, it is often impossible to prove that they derive from individuals that were hunted rather than obtained through other mechanisms (e.g. trapping). Occasionally, animal remains do exhibit evidence of hunting injuries (Leduc 2012; Noe-Nygaard 1974; Sadler 1990: 487) but it seems likely that many of the wild animals represented in the archaeological record were obtained using snares or traps, such as the 'wolf pits' noted in Scandinavian place-names (Pluskowski 2006: 20–21).

Although trapping seldom entails the same level of ritualized social performance that is associated with hunting proper, various studies (Morris 1998: 79–84; Willerslev 2007) have demonstrated that these actions are socially meaningful. Indeed recent work by Howard (2013) has highlighted that the trapping of smaller prey species is, across cultures, often the task of women and children, social groups that are frequently overlooked in the archaeological record. Interestingly, smaller prey animals are also often overlooked by zooarchaeologists in favour of the more impressive larger ungulates, such as deer and boar, thus we are unwittingly perpetuating an adult male-centric view of the past (see Howard 2013). I am ashamed to say that this chapter falls into precisely this trap, overlooking smaller animals almost entirely as my work on hunting has always focused on large prey animals. However, out of both necessity and genuine interest, it will examine zooarchaeological evidence relating to wild mammals, and to a lesser extent birds and fish, regardless of whether they were obtained by 'real' hunting, or other mechanisms.

Figure 3.1 summarizes diachronic variation in the frequency of wild mammals (shown as a percentage of the total mammalian fragment count) for zooarchaeological assemblages from southern England. Table 3.1 presents the same dataset, this time showing the percentage of assemblages in which the main wild mammals are represented. These data have been synthesized from the work of Allen (2010), Hambleton (2008), Poole (2010b), Serjeantson (2011) and Sykes (2007b), all of which used broadly comparable methods. Fewer data are available concerning the exploitation of wild birds and fish, but recent studies (Allen and Sykes 2011; Reynolds 2013; Sykes 2011) indicate that their chronological patterns closely mirror that shown in Figure 3.1, which can perhaps be used as a proxy for all wild animal exploitation.

It is important to remember that the trends apparent in these datasets are the result of human activity and cultural choices; they are not a direct reflection of the local fauna. Where wild animals are absent or present only in low frequencies, it does not necessarily mean that they were absent in the landscape, as often suggested (e.g. Jennbert 2011: 80–6). To be sure, poor representation can reflect low

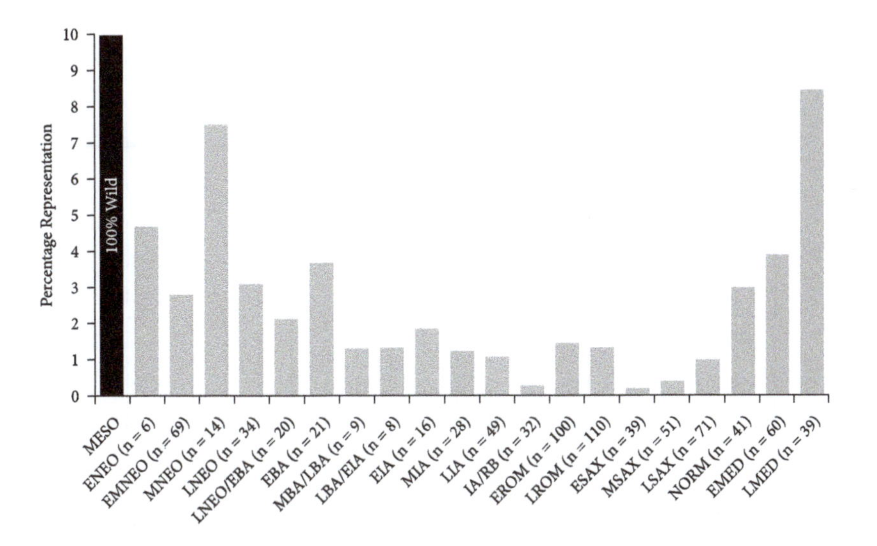

Figure 3.1 Representation of main wild mammals – red deer, roe deer, fallow deer, wild boar, aurochsen, bear and hare – according to NISP and expressed as a percentage of the total mammalian assemblage.

Sources: Allen (2010); Hambleton (2008); Poole (2010b); Serjeantson (2011) and Sykes (2007b)

Table 3.1 Diachronic variations in the representation of the main wild mammals on archaeological sites, shown in terms of the percentage of sites on which each species is represented.

Period	Red	Roe	Fallow	Wild Boar	Aurochs	Bear	Hare
Early Neolithic (7)	57.1	42.8		14.2	28.5		
Early/Middle Neolithic (69)	57.9	36.2		5.7	23.1		5.7
Middle Neolithic (n=14)	50.0	14.0			21.0		
Late Neolithic (34)	58.8	26.4		20.5	23.5	17.0	2.9
Late Neolithic/Early Bronze Age (n=20)	30.0	20.0		5.0	20.0		5.0
Early Bronze Age (n=21)	38.0	14.0		4.7	4.7		4.7
Middle/Late Bronze Age (n=9)	55.0	33.0					11.0
Late Bronze Age/Early Iron Age (n=15)	100.0	46.0		20.0			13.0
Early Iron Age (n=16)	68.0	43.0	6.0				18.7
Middle Iron Age (n=28)	53.5	28.5					39.2
Late Iron Age (n=49)	83.6	30.6		2.0			30.1
Iron Age/Roman (n=32)	75.0	40.6					40.6
Early Roman (n=100)	80.0	61.0	4.0			2.0	54.0
Late Roman (n=110)	89.0	57.0	8.0				59.0
Early Anglo-Saxon (n=39)	64.0	20.5				10.2	5.0
Middle Anglo-Saxon (n=51)	52.0	39.0		11.0			19.0
Late Anglo-Saxon (n=71)	54.0	54.0		5.0			33.8
Norman (n=41)	54.0	44.0	46.0	12.0			39.0
Early Medieval (n=60)	56.0	43.0	43.0	3.0			56.0
Later Medieval (n=39)	48.0	33.0	64.0				48.0

Sources: Allen (2010), Hambleton (2008), Poole (2010), Serjeantson (2011) and Sykes (2007b).

populations but it can equally reflect lack of exploitation by people. On the other hand, some animals may be represented archaeologically in areas where they were not present in life, their remains having been curated and transported to regions beyond their natural distribution: for instance drinking vessels made from aurochs horn were recovered from the seventh-century AD princely burial from Sutton Hoo (Norfolk), despite the fact that the aurochs had been locally extinct for several thousand years at this point (Sykes 2011: 333; see also Chapter 6, page 130). Nevertheless, if viewed as a broad-brush indicator, the data do record the changing ways people interacted with the world around them.

Hunter-Gatherers of the Mesolithic (8000–4000 years BC)

The environment and faunal spectrum of Mesolithic England were very different to those of today: species that are now completely extinct (e.g. aurochsen) or locally extirpated (e.g. elk, wolf, bear, wild boar and beaver) were probably familiar encounters for the mobile human populations that moved through the heavily wooded landscape. As a result, the experience of Mesolithic hunter-gatherers is one entirely alien to our own, made stranger still by the complete absence of domestic animals, excepting of course the dog (Figure 3.1).

Although animals in the Mesolithic world were overwhelmingly non-domestic, there is good evidence that people sought to manage them, modifying local environments to create animal-attractive browse and by selectively culling herds (Tolan-Smith 2008: 148). Under such circumstances there was considerable potential for humans and animals to become familiar with each other's patterns of behaviour and movement, perhaps even coming to recognize each other as individuals (e.g. Overton and Hamilakis 2013: 124). Certainly the presence of healed hunting injuries on Mesolithic fauna suggests that single animals were encountered repeatedly, rather than just once (Leduc 2012; Noe-Nygaard 1974). However, as was argued in Chapter 2, it seems unlikely that the relationships between Mesolithic people and wild animals were as intimate as those that developed under the umbrella of domestication. Rather it would seem that the relationship was simultaneously distant but also one whereby humans were the more dependent of the partnership, being reliant upon animals for food, clothes and shelter.

This situation may account for the beliefs and practices of many modern hunter-gatherer communities. Whilst there is no universal 'hunter-gatherer cosmology', there are a number of traits that have been identified consistently by anthropological studies of hunter-gatherers worldwide (e.g. Conneller 2011; Ingold 2000; Nadasdy 2007; Politis and Saunders 2002; Russell 2012; Zvelebil 2008). The first is that, in the general absence of 'the domestic' (in terms of both livestock and permanent settlements), the concept of 'wild' does not exist: the world in which hunter-gatherers live is just that, a single integrated realm lacking the dichotomies of wild and domestic, nature and culture that are perceived by modern Western societies. Nor is there an absolute divide between humans and animals; although both are recognized as separate categories of being, it is widely accepted that animals are essentially the same as people, possessing both souls and rational thought. Wild animals feature strongly in the origin myths of some hunter-gatherer groups, stories recounting primordial times when people were animals, or animals people (Russell 2012: 169).

The intimate and indivisible connection of hunter-gatherer communities to their environment has been summarized by Ingold (2000: 43–50) and Nadasdy (2007) as a kind of familial relationship, whereby hunter-gatherer perceive their environment as a nurturing parent and trust that it will share its resources with them, so long as they behave with respect: by killing no more than is necessary, by treating the animals' bodies with care, by sharing the meat amongst the community, and by disposing of the remains appropriately. Within this worldview, hunting is deemed to be a vital act of regeneration and, in much the same way as the phoenix rises from the ashes, the hunted animal must die in order for its soul to be released and re-clothed with flesh so that it might return to the hunter on another occasion.

Disrespectful behaviour may be punished by the Spirit Masters or Masters of Animals whose cosmic purpose is to ensure animal welfare. They are responsible for determining human success in hunting and may seek revenge on hunters or whole communities if they are deemed to have behaved badly (Ingold 2000: 115, 125; Morris 2000; see also Chapter 7). The dangerous nature of the Spirit Masters means that shamans are frequently required to mediate between human groups and these supernatural beings, and the shamans often need to adopt animal form in order to engage in dialogue. Whilst shamans may undergo complete transformation into animals, hunters frequently transform partially, employing mimesis to gain their quarry's perspective so that they might communicate better with them and encourage them to give themselves up to the hunter (Nadasdy 2007). That animals are more courted than hunted is a belief held almost universally by hunter-gatherer societies, but admittedly it is also common amongst pastoralist and farmer groups who hunt (Cartmill 1993; Willerslev 2007). Drawing upon this evidence Conneller (2004) has proposed perhaps the most interesting re-interpretations of the evidence from the Mesolithic site of Star Carr in Yorkshire, where large numbers of perforated antler frontlets were recovered from a lake-side context. Rather than invoking the usual 'functional' or 'ritual' arguments – that the frontlets were either disguises to increase hunting success or shamanistic costumes – she suggested that the deer frontlets were perceived to carry the agency of the red deer and were worn to assist the hunters' transformation into, or mimesis of, deer so that they might encourage the animal to sacrifice itself.

The possibility that similar beliefs of respect and regeneration were held in the past is suggested by the way in which animal bones were deposited during the Mesolithic: very often remains were placed in watery locations, as was the case at Star Carr (Conneller 2011: 363). When Star Carr was originally excavated in the 1940s–1950s it was understood to be a 'typical' Mesolithic site, yielding large numbers (191) of barbed antler points, twenty-one antler frontlets and an array of other artefacts. However, as excavations of other neighbouring Mesolithic sites were carried out, and knowledge of the period increased more generally, it became clear that Star Carr is far from typical. Decades of further excavation around the ancient lake have produced just one additional barbed antler point, and no other antler frontlets have been recovered from anywhere in Britain, the only other examples coming from three sites in Germany (Zander 2013). Against this backdrop Star Carr stands out as exceptional, and Conneller (2004, 2011) proposes that the site was a sacred place where objects made from animal remains could be returned to the water, perhaps so that the animals' souls might be released to be hunted again.

The same concept of water-born reincarnation may also have applied to humans, as isolated human bones have been found comingled with animal remains in aquatic contexts from the sites of Thatcham in Berkshire and Staythorpe in Nottinghamshire (Conneller 2011: 363). Certainly cross-cultural studies have shown that reincarnation is central to most hunter-gatherer ideologies: people's souls may be reborn into their own family groups or may return in animal form but always there is a continuing circulation of spirits (Mills and Slobodin 1994; Riches 2000; Vitebsky 2006). In Britain the majority of Mesolithic human remains – individual fragments, single burials, but also cemeteries – have been recovered from caves in which little other contemporary material was found (Conneller 2011: 363). The deposition of human remains in such isolated locations indicates that some areas were considered places of the dead and there may have been a concept of an 'underworld' (Zvelebil 2008: 43).

The evidence provided from deposits of human and animal remains offers some indication of the way in which Mesolithic landscapes were conceptualized. In common with modern hunter-gatherers, the Mesolithic populations must have had a detailed knowledge of their environment, their mobile lifestyle allowing them to form interpersonal relationships with all aspects of their world and 'stay in touch' with the communities and individuals (e.g. animals, plants and spirits) that lived there. Whilst there is some suggestion that human populations continued to be fairly mobile in the early Neolithic period (Serjeantson 2011: 64), Bogaard's (2012) work on archaeobotanical material from northern Europe indicates the opposite: rather than a system of shifting agriculture, people were investing considerable time and effort cultivating single plots. This change in lifestyle – the new attention given to plant and animal husbandry, the gradual rise of permanent settlement and associated reduced mobility – must have imposed limitations on the degree to which people could continue to engage with the wider environment and may account for the changes that we see in the zooarchaeological record of these early farming communities.

Early Farming Communities of the Neolithic (c. 4000–2500 BC) and Bronze Age (c. 2500–700 BC)

In stark contrast to the Mesolithic period, zooarchaeological assemblages from Neolithic and Bronze Age sites in southern England are generally characterized by 'the domestic', the remains of cattle, sheep and pigs making up 90 to 98 per cent of most faunal assemblages (Figure 3.1). Less apparent are those wild resources that must surely have been plentiful within the landscape. For although elk (*Alces alces*) populations appear to have declined severely during the Mesolithic (Kitchener 2010), there is little evidence to suggest that other species suffered similarly and populations of aurochs (*Bos primigenius*), red deer (*Cervus elaphus*) and roe deer (*Capreolus capreolus*), wild boar, bear (*Ursus arctos*), wolf, lynx (*Lynx lynx*) and wild cat (*Felis silvestris*) presumably continued to thrive. Certainly large herds of red deer must have been locally available to provide the considerable quantities of antler that were used to construct, and were deposited within, the many monuments and mines of the Neolithic period (Serjeantson 2011). The fact that the majority of these antlers are shed suggests that deer were seldom hunted but that considerable effort went into the collection of their antlers. Indeed, Fletcher (2011: 28) has argued that red deer were actively managed for their antlers, rather than being viewed as a source of meat.

Recent reviews of the zooarchaeological data (Pollard 2006; Schulting 2013: 316; Serjeantson 2011) have confirmed that wild animals were not extensively exploited for food during the Neolithic period. This impression is reinforced by evidence from stable isotope analysis (Figure 3.2), which suggests a rapid shift away from the consumption of marine resources between the Late Mesolithic and Early Neolithic periods (Richards *et al.* 2003; Richards and Schulting 2006).

Richards and Schulting (2006) have argued that the pattern reflects the emergence of a cultural taboo over the consumption of fish and other marine

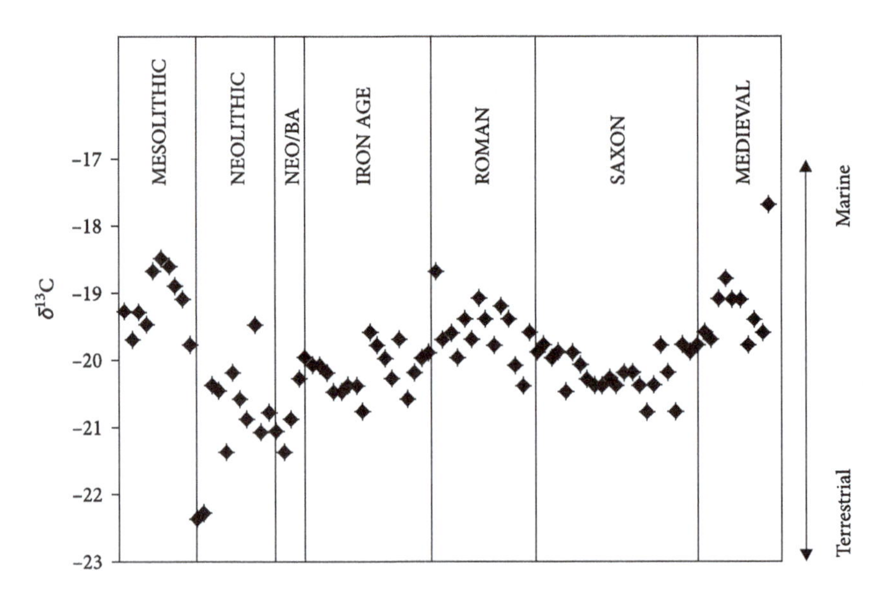

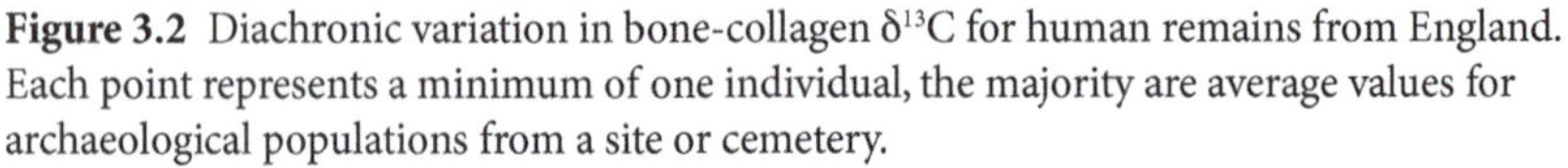
Figure 3.2 Diachronic variation in bone-collagen $\delta^{13}C$ for human remains from England. Each point represents a minimum of one individual, the majority are average values for archaeological populations from a site or cemetery.

Sources: Hedges *et al.* (2008); Reynolds (2013); Lightfoot *et al.* (2009); Stevens *et al.* (2010, 2012).

resources. This interpretation has not been accepted without criticism (e.g. Milner *et al.* 2006); however, for England at least, it does seem likely that marine resources and probably all other wild animals were quickly eschewed as regular foodstuffs from the beginning of the Neolithic, a situation that appears to have endured for almost four millennia (Figure 3.2).

Why people moved away from the procurement and consumption of wild animals will, no doubt, continue to be a subject of debate because there are many possible explanations. It may be that as mobility reduced in favour of sedentism, people 'lost touch' with areas beyond their immediate day-to-day surroundings and, as the known world contracted, the areas beyond settlements came to be perceived as the dangerous dwellings of ancestors and spirits (Ingold 2000: 43). Under such circumstances, people may have chosen not to traverse these unknown landscapes.

Alternatively, the new 'wilderness' may have hosted regular hunting trips, the evidence for which has been lost to the archaeological record because the remains of wild animals were deposited away from centres of human activity. Some support for this suggestion is provided by several finds of butchered aurochsen, all of which were deposited in pits not obviously associated with settlements (Serjeantson 2011: 79, 93). An important example of isolated deposits of wild remains is the Coneybury Anomaly in Wiltshire, which Maltby (1990) found to contain an exceptionally high frequency of roe deer bones but also the remains of red deer, beaver (*Castor fiber*) and trout (*Salmo trutta*). These faunal remains were deposited with a large ceramic assemblage, at least forty pots ranging from small cups to serving vessels (Whittle 1996: 234). This has prompted the suggestion that the deposit resulted from an episode of feasting (Exon *et al.* 2000: 31), although Serjeantson (2011: 47) prefers the Coneybury Anomaly as representing a specialized hunting camp. There is no reason that these two hypotheses should be mutually exclusive, as there is considerable cross-cultural evidence to suggest a strong connection between hunting and feasting (Steel 2004: 292; Sykes in press; Wright 2004). Both are often male-only activities and it is not unreasonable to assume that the same was true in the Neolithic period. To take the interpretation further: given the remote location of the Coneybury Anomaly, and other Neolithic deposits of wild animals, it is tempting to suggest that hunting may have formed part of a maturation ritual, whereby young men were required to spend a period of time living in the wilderness before being reintroduced to society as men. Such coming-of-age rites are common amongst many traditional societies today (Morris 1998: 61–119) and have been intimated for Prehistoric societies across Europe (Koehl 1986; Sykes in press). If this were also the case in Neolithic England it would suggest that hunting was a 'special' activity, rather than part of daily practice.

The possibility that wild animals were seen as 'special' in the Neolithic period is perhaps indicated by the fact that, although they are poorly represented in domestic assemblages, they are found in higher frequencies in deposits associated with monumental and funerary structures. Studies by Pollard (2006) and Thomas and McFadyen (2010) have shown that, in earlier Neolithic assemblages, the remains of wild and domestic animals were not obviously subject to differential treatment – both were frequently mixed together, often in the same deposits as human remains. Pollard (2006: 139) takes this lack of symbolic distinction as evidence that wild animals had a 'diminished value in the context of social practice'. However, this argument only stands if domestic animals were also not seen as special, which seems improbable given that they were recent introductions to England at this point and, as is the case with most exotica (see Chapter 4), were evidently central to human society and the negotiation of social relations (Pollard 2006; Ray and Thomas 2003).

Pollard's (2006: 140) proposal that both the general lack of wild animals and their association with funerary contexts might be attributed to Neolithic cosmologies is more persuasive. The mixing of human and animal remains indicates a worldview where, as in the Mesolithic, no definitive human–animal separation was perceived. The idea that humans and animals, nature and culture, life and death were indivisible is also suggested by the mortuary traditions of the time. Whilst some human remains were deposited within funerary structures, excarnation or exposure were apparently more common funerary rites. These practices would have seen human bodies returned to the elements – or the ancestral realm – via decay and animal scavenging. If, as many anthropological studies suggest, the wilderness was seen as the spirit world or wild animals as ancestral incarnations, it is to be expected that cultural taboos may arise around hunting and the exploitation of wild resources, as Pollard (2006) has argued for the Neolithic.

The situation for the Late Neolithic/Bronze Age seems slightly different. Figure 3.1 indicates that the representation of wild animals declines further. This trend is seemingly confirmed by Table 3.1, which shows that, of all the periods under consideration, the Late Neolithic/Bronze Age and Bronze Age periods have the lowest percentage of assemblages containing wild animal remains: red deer, usually the best represented wild species, are found on 55 per cent of Middle to Late Bronze Age sites and just 30 per cent of those dating to the Late Neolithic/Early Bronze Age. Indeed, the virtual absence of wild animals is, in general, a trait of Bronze Age assemblages across Europe (Harding 2000: 136, 143). This shift could feasibly represent an intensification of a Neolithic wild resources taboo but other explanations seem equally possible, especially when the zooarchaeological data are considered in context.

The transition from the Late Neolithic to Bronze Age was marked by many cultural changes. From the very beginnings of the Early Bronze Age, there is evidence for a dramatic increase in long-distance international trade, with the arrival of the 'Beaker Package', which brought the domestic horse (Bendrey 2010), new archery gear, exotic objects of bronze and other materials, innovative ceramic styles (Brück 2006a), as well as a shift in funerary practices (Bradley and Fraser 2010). These introductions were accompanied by changes in the cultural landscape. For instance, in the second half of the Early Bronze Age (around 1850–1500 BC) people began to colonize the heathlands of southern England, areas which had not been occupied during the Neolithic or Late Neolithic/Early Bronze Age (Bradley and Fraser 2010). By the end of the Early Bronze and through the course of the Middle Bronze Age, the landscape was transformed completely as field systems, droveways and stock handling structures were laid out, suggesting a new emphasis on large-scale, long-distance animal herding (Yates 2007). Together these changes indicate a fundamental shift in the way that people were engaging with, moving through and perceiving their world. Indeed, it seems possible that they both reflect, and were instigated by, a recalibration of worldview brought about by increased international trade and human migration (stable isotope studies suggest that there was an explosion of human migration during the Late Neolithic and Bronze Age – e.g. Price *et al.* 2004; Evans personal communication).

My suggestion is that, as people travelled further and were introduced to the new 'magical' material of bronze, which had to be conjured through the smelting of rock ores, the wilderness and wild animals may have lost some of their status as people came to view the 'exotic' as the new and powerful outer sphere (Helms 1993). The idea that the exotic replaced the wilderness as the sacred realm is indicated by the shift in burial practices: away from exposure or excarnation, initially towards inhumation and then cremations, the latter two generally associated with burial mounds (Bradley and Fraser 2010). Although excarnation continued to some extent (Brück 2006b: 309), the rise of inhumation – where bodies were deeply buried and sometimes sealed further with mounds – suggests that people were beginning to believe that it was inappropriate for cadavers to be consumed by wild animals. Furthermore, as cremation became the common mortuary practice, human bodies were increasingly treated in ways that mimicked the transformation of bronze smelting, with cadavers being burnt and pounded like metal ores: Brück (2006a: 86) has highlighted that the widespread use of bronze and the adoption of cremation are entirely synchronous.

A decline in the cultural importance of wild animals is perhaps to be expected in a period when pastoralism, and in particular cattle, became central to society: some modern pastoral societies do view wild animals with a certain level of disdain (Russell 2012: 170; Abbink 1993: 709). It may be that in the Late Neolithic and Bronze Age of England hunting was, indeed, an activity that took place only at times of need. Certainly the representation of red deer and roe deer is exceptionally low when viewed against all other periods (Table 3.1). However, the one argument against human ambivalence to wild animals in this period is based upon the aurochs, the large and dangerous wild cattle. Table 3.1 shows that aurochsen are represented in a considerable percentage of assemblages dating to the Neolithic. This high frequency continues into the Late Neolithic/Bronze Age, when aurochsen become one of the best represented wild species, found on a comparable number of sites and in comparable relative frequencies to red deer (Serjeantson 2011: 38). According to Table 3.1, the representation of the aurochs drops dramatically during the Early Bronze Age, but the remains of wild cattle do continue to be found, particularly associated with the burials of elite males – many of whom were also interred with archery kits (Cotton *et al.* 2006; Davis and Payne 1993; Towers *et al.* 2010). Conclusive evidence that aurochsen were taken by bow and arrow is provided by the Early Bronze Age skeleton from Hillingdon (London), which was recovered from a pit, its remains closely associated with six tang and barbed arrow heads (Serjeantson 2011: 45).

The connection between archery and elite males points to a situation where wild cattle were bow-hunted as a display of power, martial ability and, presumably, masculine identity (Ray and Thomas 2003: 42). This would be consistent with the situation seen across Bronze Age Europe, where the rise of warrior cultures saw hunting become important (albeit ideologically rather than actually) for the negotiation of power and gender roles (Hamilakis 2003; Treherne 1995). Whilst the low frequency of wild animals in Bronze Age assemblages indicates that hunting was an occasional activity, it clearly took place at a level sufficient enough to cause the local extinction of the aurochs (Table 3.1 and see Legge 2010).

Iron Age (c. 750 BC–AD 43) and Roman Periods (AD 43–410)

The demise of the aurochs was not the only decline that occurred towards the end of the Bronze Age. Evidence suggests that the period saw climatic deterioration and a collapse of international trade in exotic goods, with the bronze networks falling out of use by *c.* 600BC (Henderson 2007: 95, 116). It may be no coincidence that the end of international trade apparently heralded a return to a more parochial worldview whereby the wilderness, rather than the exotic, again became the sacred realm. This is indicated by the widespread reversion to excarnation/exposure as the main funerary practice, evidenced by the general absence of formal burial and the increased representation of disarticulated human remains, including some that show clear traces of carnivore gnawing (Craig *et al.* 2005; Redfern 2008; Smith 2006). According to Serjeantson and Morris (2011: 101) scavenging birds such as corvids and buzzards (*Buteo buteo*) were important agents of the excarnation process, perhaps explaining why these species were so frequently the subject of so called 'special deposits' (complete/partial animal burials also known also as Associated Bone Groups or 'ABGs') that are frequently found associated with fragments of human bone (see Chapter 6).

Figure 3.1 shows that wild mammals continue to be rare in the Iron Age archaeological record but, importantly, they are better represented in ABGs than they are in other contexts: it should be noted however that wild animals are still rare, most ABGs comprising domestic animals (see Morris 2008a, 2008b). Interestingly, Van der Veen (2008) has suggested that a similar situation is true of wild plant remains – that they are rare in domestic deposits but more common in ritual deposits. Based on the disparity in the representation of wild animals, Hill (1995: 64, 104) proposed that the hunting and consumption of game were proscribed, only being undertaken on rare occasions of feasting and sacrifice. Such a scenario would certainly account for Caesar's account of the Britons, where he stated with incredulity that 'hare, fowl and geese they think it unlawful to eat' (V.12 – trans. Handford 1982: 111). The fact that Caesar singled out the hare (*Lepus* sp.) for special mention is telling, especially when the data in Table 3.1 are examined – here it can be seen that hare remains become well represented for the first time, being present on 13 per cent of Late Bronze Age/Early Iron Age assemblages and almost 40 per cent of those dating to the Middle Iron Age. A recent interdisciplinary study by Crummy (2013) has highlighted that there is little evidence these animals were eaten and it seems more likely that their abundance as ABGs from Iron Age southern England, and much of Europe, suggests the presence of a Celtic hare-deity (Boyle 1973: 315; Crummy 2013; Green 1992: 51–8, 125. See also Chapter 4).

The possibility that hares, and other wild things, may have been viewed as sacred during the Iron Age is indicated not only by the apparently ritual treatment of wild animals but also by the iconography of the period. Human–animal hybrids are frequently depicted (e.g. Aldhouse-Green 2004: 150, 170–8) which resonates with the cosmologies of many traditional societies, who believe that whilst in the wilderness humans may take the form of animals or exchange body parts with the spirits who dwell there (Ingold 2000: 84). Perhaps the most famous human–animal hybrid depiction is illustrated on the large silver cauldron recovered near Gundestrup, Denmark. In one panel an antlered man, believed to represent Cernunnos the Celtic god of animals, is sitting cross-legged amongst a menagerie of wild animals, the location presumably the wilderness (see Figure 3.3a; Green 1992: 146).

Creighton (1995: 297) has suggested that the wilderness may have been home to Druids, the shaman-like individuals of the Iron Age who negotiated between humans and the spirit world, in much the same way as is depicted upon the Gundestrup Cauldron. To support his case that the wilderness was perceived as a boundary or liminal space during the Iron Age, Creighton (1995: 297–8) maps the votive offerings of coins and metalwork to show that they were frequently found at the edges of bogs, lakes, rivers, estuaries and the sea shore, areas where evidence for Iron Age settlement and day-to-day activities is limited (Willis 2007: 115).

Water clearly played an important role in religious activities throughout Prehistory, and Madgwick (2008) prefers to see water burial, as opposed to excarnation, as the dominant mortuary practice of the Early and Middle Iron Age. This proposal accords well with the finds of human skulls in the Thames and other rivers perhaps suggesting that human bodies were treated in the same fashion as the many metal objects also deposited in watery locations (Bradley and Gordon 1988; Creighton 1995). The idea that watery places were viewed as sacred, or polluted due to their association with corpse disposal, may also help to explain why people in southern England continued to avoid eating fish and other animals from the water. This is indicated not only by the absence of fish bones and shellfish in archaeological deposits (Dobney and Ervynck 2007; Serjeantson *et al.* 1984; Willis 2007) but also the isotopic evidence for Iron Age populations, which continues to show low $\delta^{13}C$ values more typical of a terrestrial-only diet (Figure 3.2; Reynolds 2013).

In some respects the Later Iron Age and Roman attitudes to wild animals and the wilderness were similar to those of the Early and Middle Iron Age: there is evidence that watery places and their animals continued to hold religious significance (Willis 2007), that votive offerings were still made at boundary locations (Rogers 2007), and that there was some continuity in the ritual deposition of wild animals (Morris 2008a). However, it is interesting to note that exotic animals, such as chickens, begin to be found in higher frequencies within 'ritual' deposits (e.g. King 2005; Morris 2008b; and see Chapter 4) whilst wild animals become better represented in domestic contexts. There would seem to be a metaphorical link between the 'wilderness' and the 'exotic' – both being beyond the domestic and perceived as 'outside'. My suggestion is that, as in the Bronze Age, the increased importation of exotic species and other goods during the Late Iron Age and Roman periods may have expanded mental geographies, the exotic realms beyond the known limits of the world coming to replace the wilderness as the new sacred spheres (Sykes 2010a).

If the boundaries to divine spheres were shifting, and the traditional wilderness was losing its sacred status, this may explain a number of the changes that occurred during the course of the Iron Age/Roman transition. First, there is clear evidence that coastal areas and fenlands began to be settled, suggesting that it was increasingly acceptable for humans to inhabit these spaces, as well as being physically possible due to a combination of marine regression and the construction of drainage networks (Willis 2007: 119). Second, the Late Iron Age and Roman periods saw funerary practices change once again to a situation where human bodies were either buried or cremated, ensuring that they could not be consumed by wild animals, a prospect that was abhorrent to the Roman belief system (Toynbee 1971: 43). Third, iconographic representation changes somewhat, with human–animal hybrids (Figure 3.3a) becoming less common: rather than being mixed together with wild animals, Roman gods and goddesses are depicted in human form but with wild animal familiars (Figure 3.3b) – suggesting that humans and animals were coming to be viewed as very separate entities. According to Gilhus (2006: 79) discussions about human–animal transformations and hybridizations within classical Greek and Roman texts were not portrayed in a positive light.

By contrast to the Iron Age, where evidence suggests a deep reverence for the wilderness, the Roman data indicate that hunting was less a religious experience and more a maker of elite identity. According to Dunbabin (2004: 146) hunting scenes emerged as a popular motif from the third century onwards. This was at approximately the same point that the possession of productive estates, hounds, horses and leisure time was coming to be viewed as the defining traits of elite identity – hunting serving as the nexus through which all could be communicated. This is also indicated by classical literature (Anderson 1985).

It is often suggested that the zooarchaeological record does not reflect the passion for hunting suggested by Late Iron Age and Roman texts and iconography (Cool 2006: 111–18) and, looking at Figure 3.1, it is easy to see why. Overall, the representation of wild animals does not increase dramatically but this is where context becomes important: those who have studied the zooarchaeological evidence in detail have demonstrated a clear rise in in the utilization of wild resources – in particular deer, hare, wildfowl, fish and crustacea – as a source of food, particularly on elite settlements such as large villas and military centres (Allen 2010). A recent reanalysis of the assemblage from Fishbourne Roman Palace (Sussex) identified a raft of wild mammals and birds, including a large number of cranes and other waterfowl that were presumably caught on the nearby marshlands (Allen and Sykes 2011). At this site, and some other Roman settlements,

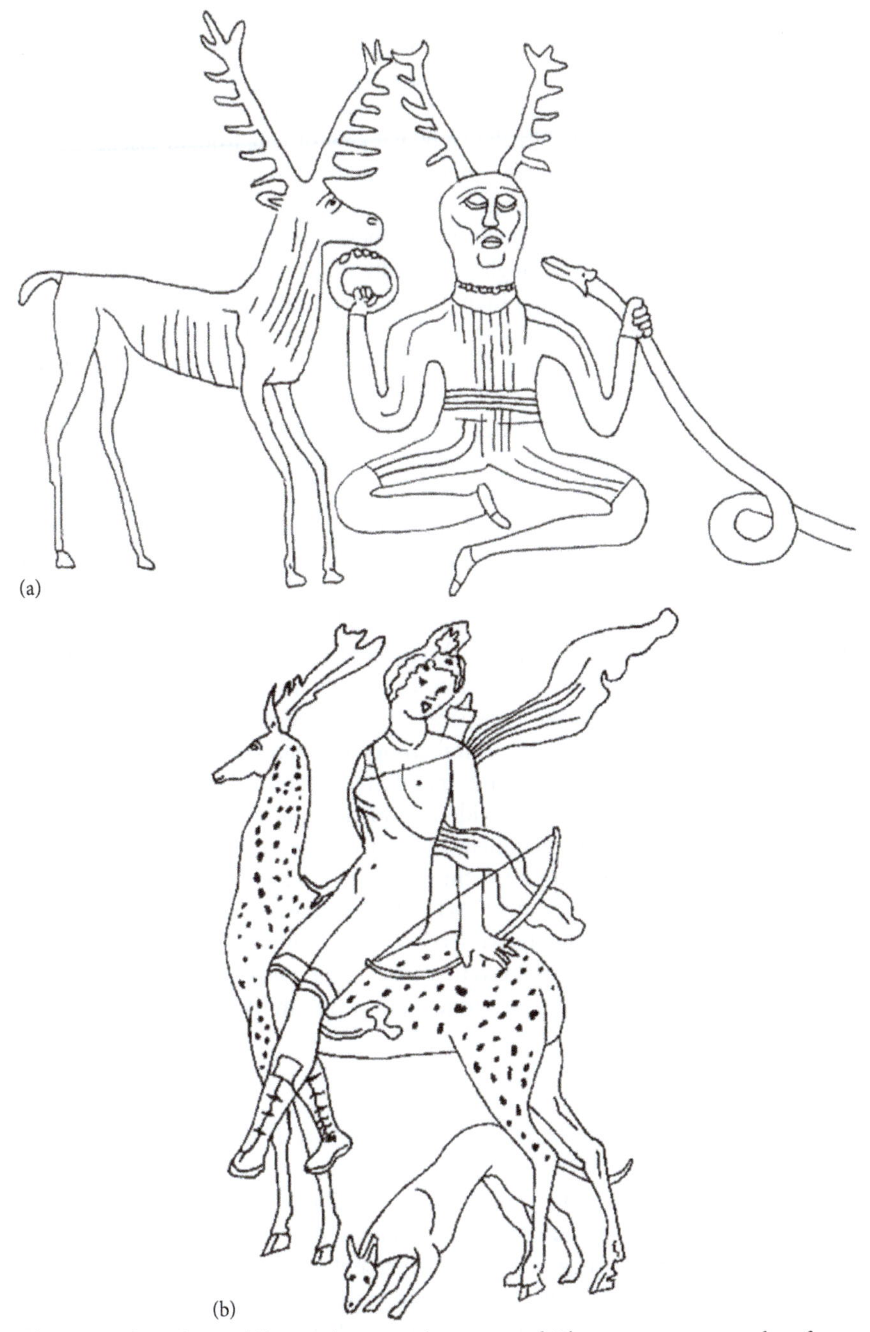

Figure 3.3 Iron Age and Roman iconography compared. There are more examples of hybridized deities in the Iron Age (such as is depicted on the Gunderstrip Cauldron – (a)) whereas animals are more often shown accompanying Roman deities, as is shown above with Diana riding a fallow deer (b).

Source: David Taylor

there is evidence that parks were established in the landscape to house exotic fallow deer, *Dama dama* (see Table 3.1; Chapters 4 and 5; Sykes *et al.* 2006; Sykes 2010a). The presence of *leporaria* (enclosures specifically for hares) may explain the exceptionally high frequency of hares seen in the Roman period: according to Table 3.1 they are represented in almost 60 per cent of Late Roman assemblages, a frequency that was never again surpassed and contrasts markedly with the evidence for the succeeding periods.

Fish also began to be maintained in ornamental ponds, or *piscinae*. The first-century BC author Varro, although admittedly not writing about Britain, stated that they 'appeal to the eye more than to the purse, and exhaust the pouch of the owner rather than fill it' (*De Re Rustica* III.XVII.2–4 trans. Hooper and Ash 1979). Physical evidence for *piscinae* is scarce in England but ponds have been tentatively identified at Fishbourne Roman Palace where fish remains have also been recovered in considerable quantities (Allen and Sykes 2011). Locker (2007) has undertaken a detailed review of the zooarchaeological evidence for fish, demonstrating a clear increase in their consumption during the Late Iron Age and Roman period. This pattern is confirmed by isotope analyses of human remains, which show a general trend towards increased $\delta^{13}C$ values indicative of a marine diet (Figure 3.2; Reynolds 2013).

Fish and other wild animals were seemingly still considered to be 'special' (e.g. Morris 2008a: 152; Willis 2007: 114), so their sacred association may have remained, but their procurement and consumption was clearly no longer prohibited. This is a key point which would seem to separate the ideologies of the earlier Iron Age and Roman society. Whilst both cultural groups saw nature and the wilderness as sacred, their beliefs appear to have manifested themselves in different ways. Archaeological and iconographic evidence indicate that the Iron Age population negotiated with the world around them, their cosmology reflected by the avoidance of wild resources and the ritual treatment of the animals derived from 'outside' (Aldhouse-Green 2004; Green 1992: 241). The Romans, on the other hand, saw it as their spiritual duty to bring the wilderness to order, investing their efforts in the paradox of domesticating the wild so that they might dwell in the manner of their gods in close proximity to, but separate from, wild animals (Beagon 1996: 299; Coates 1998: 27; Purcell 1987: 201).

Early and Middle Anglo-Saxon Period (AD 410–c. 850)

The withdrawal of the Roman Empire from Britain in AD 410 took with it the hunting, fowling, and fish-eating culture that had developed over the four centuries of occupation. Table 3.1 suggests that the representation of hare falls dramatically, the taxon being identified in just 5 per cent of Early Anglo-Saxon assemblages. The presence of red and roe deer is also substantially lower in the Early Anglo-Saxon period compared to their representation in Roman assemblages and there is no evidence to suggest that fallow deer populations endured into the post-Roman period (Table 3.1; Sykes and Carden 2011).

The overall scarcity of wild animals in archaeological assemblages is demonstrated well by Figure 3.1, which suggests that wild mammals are found in lower frequencies in the Early Anglo-Saxon assemblages than in those from any other period. Wild birds and fish are equally scarce on Early Anglo-Saxon sites even where sieving was undertaken (Sykes 2011). The general dearth of fish bones in Early Anglo-Saxon deposits, together with the results from isotopic analysis of human remains, which revert to the low $\delta^{13}C$ values more consistent with the pre-Roman terrestrial diet (Figure 3.2), suggest that the Early Anglo-Saxon population consumed few animals from the water.

The poor representation of wild animals in fifth- to seventh-century assemblages promotes an impression of a landscape devoid of wildlife and it is possible that over-hunting in the Roman period, combined with woodland clearance, did reduce populations of wild animals. However, the evidence from non-anthropogenic deposits paints a different picture. Radiocarbon dating of the animal remains from Kinsey Cave in Yorkshire demonstrated that both brown bear and lynx were living in Early Anglo-Saxon England: a lynx femur from this site returned a calibrated date of AD 425 to AD 600 (Hetherington *et al.* 2006) and a bear vertebra was dated to the fifth/early sixth century (Hammon 2010). Old English place-names also suggest that the Anglo-Saxon landscape was populated with a wider variety of wild species than exist in today's landscape: wolves, bears, lynx, beavers, white-tailed eagles (*Haliaeetus albicilla*) and cranes (*Gus grus*) abound in place-names dating to the Middle and Late Anglo-Saxon period (Aybes and Yalden 1995; Gelling 1987; Hammon 2010; Yalden 1999: 11).

Place-names are a useful source of information, providing a snapshot of contemporary landscape perception. Semple's (2010) study of pre-Christian place-names has shown that wild and natural places – fissures, caves, hollows, hilltops, rivers, pools, springs, wetlands and moors – were believed to have sacred or supernatural associations: they were the dwellings of elves, goblins and terrifying monsters, such as *Beowulf*'s Grendel who lurked at the bottom of a mere. Such associations may explain why people seemingly avoided these locations and certainly derived little sustenance from hunting, fowling, and fishing. As in many of the preceding periods, watery locations continued to be the foci of votive deposits; considerable quantities of weapons, jewellery and tools have been recovered from rivers, particularly in areas near to crossing points (Lund 2010). Taken together the evidence suggests very strongly that the Early Anglo-Saxon natural world was perceived as both a sacred place and a scary place where both wild and supernatural creatures lived.

One needs only to look at the material culture of the period to realize that wild animals carried an importance far beyond economics and diet. While wild creatures figure small in animal bone assemblages from domestic sites, they are more prominent in the iconographic and artefactual record (Dickinson 2005; Hawkes 1997; Pluskowski 2010). This is particularly the case for fish, predatory birds and dangerous mammals (notably bears and wild boar), which is interesting given that Table 3.1 shows both bears and wild boar to be well represented in the Early and Middle Anglo-Saxon period. Finds of bear phalanges in human cremation urns led Bond (1996) to conclude that furs played a role in Early Anglo-Saxon funerary rites. Elsewhere, bear claws appear to have been used as amulets, and it is as amulets that wild animals are perhaps best represented in this period: individual bear claws, eagle talons, and boar, wolf, and even beaver teeth have been recovered from funerary contexts (Crabtree 1991b; Lucy 2000; Meaney 1981). Where it is possible to age and sex the human skeletons, these amulets are often found with women and children, perhaps suggesting an association with fertility or protection (Meaney 1981: 134).

There is considerable debate about how dramatically the return of Christianity in the seventh century changed worldviews but it certainly affected the range of evidence available to reconstruct Middle Anglo-Saxon attitudes towards wild animals. With the end of grave-goods deposition and the wearing of anything but Christian amulets, we lack the sources of information that were so useful for the preceding period (Meaney 1992: 116). The zooarchaeological evidence does suggest that perceptions of wild animals began to change: although the remains of wild mammals, birds and fish continue to be present only in low frequencies (Table 3.1, Figure 3.1 and Figure 3.2), there is a clear trend towards increased abundance of butchered and burnt deer bones amongst food remains, indicating that these animals were being eaten rather than simply transformed into material culture (Sykes 2011). This shift could feasibly reflect the cosmological changes brought by the conversion to Christianity; for while Pagan beliefs are often 'zoocentric', with animals seen as having power equal to or in excess of humans, Christian paradigms relegated animals in the Chain of Being, suggesting that their existence was solely for human benefit (Pluskowski 2006, 2010). If Christian doctrine made hunting more acceptable, it also became an activity that appears to have brought communities together in the pursuit and consumption of wild animals (Sykes 2010b; see Chapter 8). Although all members of Anglo-Saxon society appear to have participated in hunting and received shares of the meat, there is some evidence that wild animals are marginally better represented on high-status settlements (Sykes 2011): a trend that was set to continue into the later Anglo-Saxon period.

Late Anglo-Saxon and Norman Periods (c. AD 850–1150)

Figure 3.1 shows that the Late Anglo-Saxon period saw increased representation of wild fauna. This is, in part, related to the development of craft industries, notably bone/antler working (MacGregor 1991) and the fur trade (Fairnell 2003), which were becoming established in newly forming towns. Evidence that wild animals were beginning to be viewed as commercial food resources, rather than sacred creatures, is provided by the late tenth-century school-book know as *Aelfric's Colloquy*, which gives details about the lives of professional huntsmen, fowlers, and fishers. For instance, Aelfric's fisherman explains how he sold his catch to the urban population (Garmonsway 1939: 26–30), and a rise in fish consumption around AD 1000 is indicated not only by the zooarchaeological record (Barrett *et al.* 2004) but also by isotope studies of human remains, which indicate a contemporary increase in the consumption of marine resources (Figure 3.2).

Motivation for the rise of the fishing industry is unclear and different scholars have proposed various hypotheses to account for the change: the introduction of new fishing technologies (driftnets or long line and hooks), climate change, Viking influence, widespread adoption of Christian fasting practices, and commercial revolution have all been suggested (Barrett *et al.* 2004; Serjeantson and Woolgar 2006; Tsurushima 2007). However, Reynolds (2013) has shown that high-status sites in southern England provide the earliest evidence for widespread consumption of very large cod, suggesting that the elite may have driven the fashion, perhaps 'hunting' large sea fish as a mechanism for displaying seamanship and authority. Certainly Gardiner's (1997) historical and archaeological work on cetacean exploitation concluded that whales and porpoises were viewed as elite property from the early eleventh century, their ownership and consumption becoming an element of competitive display amongst the Late Anglo-Saxon aristocracy.

The status associations attached to the hunting of land mammals are clear from documentary sources. Aelfric Bata's colloquies, written at the start of the eleventh century, referred to hunting as an activity of 'kings and great men' and Asser's *Vita Aelfredi* (*c.*893) mentioned hunting, alongside reading and warfare, as part of princely education (Marvin 2006). Even where hunting was undertaken by professional huntsmen, rather than kings or lords themselves, these hunt servants ranked socially higher than other labourers: Aelfric's huntsman held the first seat in the king's hall and was well rewarded for his service (Garmonsway 1939: 83). Aelfric's hunter provides a considerable amount of information about his favoured quarry and the methods he uses to catch them. He makes no mention of bows and arrows, which is perhaps unsurprising given that there is little historical, iconographic, or archaeological evidence for their use in Late Anglo-Saxon England (Lewis 2007: 113). Instead, he states that, using dogs, he drove wild animals – wild boar, stags, roe deer, and hare – into nets and killed them once they became ensnared (Garmonsway 1939: 56–60).

This 'drive' method is believed to have been the standard pre-Conquest hunting technique, employed by both professionals and elite alike (Cummins 1988; Gilbert 1979). It would have been a highly effective technique for obtaining large quantities of game in a single event, although to be successful it would have required the participation of many people. It may be for this reason that the Domesday Book mentions that citizens of Hereford, Shrewsbury, and Berkshire were obliged legally to act as game-beaters for their lords (Loyn 1970: 366). It is often suggested that hunting and venison consumption became linked to the aristocracy only after 1066, when William the Duke of Normandy (France) succeeded in conquering the country and much of Britain. However, the evidence presented above makes it clear that the Normans would have been familiar with the hunting institutions already in place in Late Anglo-Saxon England.

Nevertheless, 1066 does seem to mark a point of dramatic change in wild animal exploitation, Figure 3.1 showing that the representation of wild animals increases substantially from this point onwards. More detailed studies of the evidence (Sykes 2005, 2006a, 2007b) have shown that this post-Conquest increase in wild animal frequencies is due almost entirely to the efforts of the aristocracy. On lower-status settlements, the representation of wild animals actually drops, hinting at the type of unequal access to wild resources that would have accompanied the Norman introduction of forest law, under which unlicensed use of forest resources was punishable by imprisonment or maiming. Forest law blurred geographic definition of wild and domestic landscapes: no longer was there a 'wilderness' separate from the domestic sphere, instead the Normans superimposed their own hunting landscape over the existing domestic one. By restricting hunting rights to the elite, the 'wild' became associated with high social status whereas 'domestic' equated to the lower classes. Essentially forest law placed game animals in the same macro-category as the aristocracy, a category above the peasantry who were now essentially classified together with domestic animals.

The higher frequencies of wild mammals found in many Norman period assemblages enables a more detailed investigation of the changes in hunting traditions that resulted from the Conquest. For instance, Figure 3.4a shows anatomical data for deer from Late Anglo-Saxon elite sites against those from elite Norman sites (Figure 3.4b) and the patterns are very different: whereas all parts of the deer skeleton are well represented in the Late Anglo-Saxon assemblages, those from Norman sites show an overwhelming abundance of lower hind limb bones, a dearth of elements from the upper forelimb and an almost complete lack of the pelvis. On multi-period sites, where dating permits, it can be seen that deer body part patterns change dramatically before the end of the eleventh century. I propose that the bizarre skeletal patterns that characterize Norman elite sites reflect the origins of the 'unmaking' procedure, a ceremony associated with a new kind of hunting – *par force* hunting – that is well documented for the later medieval period

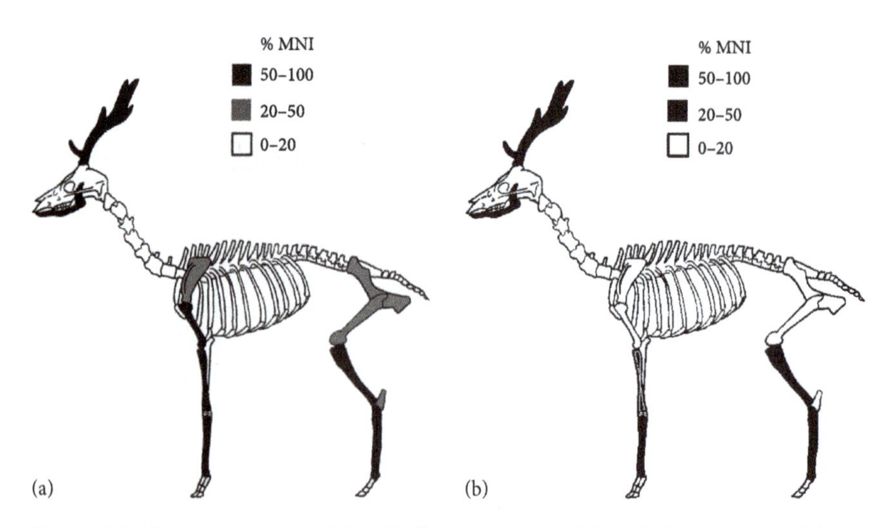

Figure 3.4 The representation of deer body parts recovered from high-status assemblages dating to (a) the Late Anglo-Saxon period and (b) the Norman period.

(Almond 2003; Cummins 1988). The *chasse par force* was fundamentally different from the drive techniques of the pre-Conquest period. It was a wide-ranging hunt of day-long duration in which a single deer was stalked, killed and 'unmade' (skinned, disemboweled and butchered) in a ritualized and formulaic manner. By comparison with the drive, *par force* hunting was an inefficient means of obtaining venison and it must be assumed that sport and, more particularly, social display were the prime motivations for this form of hunting.

The unmaking of the deer took place at the kill-spot and was the pinnacle of the hunt where the hunters were able to display their skill and knowledge of the complex unmaking procedure. Instructions vary considerably between hunting manuals (for examples see Cummins 1988: 180) but in general the texts suggest that, in the case of the stag, its testicles and penis were first removed and hung on a forked stick which was used to collect together all the prized organs and titbits. The animal was then skinned, first split from the chin down to the genitals and out to each leg before being flayed down to the spine. At this point the feet were removed from the carcass but left attached to the skin, which was now used as a convenient blanket upon which to undertake the butchery. The shoulders and haunches were removed and then the rest of the carcass de-fleshed with the meat and antlers being carried home in the skin, presumably using the feet as handles. Certain parts of the carcass did not, however, return with the lord to his abode: the pelvis was left at the kill-site as an offering to the raven, whilst the right and left shoulders were presented to the best huntsmen and the forester or parker as his fee. It must surely be the gifting of these body parts that account for the under-representation of the shoulder bones and pelvis within the assemblages from high-status sites.

By the mid-twelfth century, knowledge of these unmaking rituals and, in particular, the French terminology surrounding them was deemed to be a mark of nobility, a fact condemned by the contemporary moralist John of Salisbury who wrote that 'the scholarship of the aristocracy consists in hunting jargon' (*Policraticus* 1.4, I.23 trans. Keats-Rohan 1993). That the language of the hunt was French and that deer body part patterns suggestive of the unmaking appear shortly after 1066, strongly implicates the Normans as importers of the *chasse par force*. Elsewhere I have argued that hunting, and in particular the introduction of the unmaking rituals, provided the Norman elite with a mechanism through which they could project everything they wished to emphasize about themselves: their 'Frenchness', their social superiority, their masculinity and martial supremacy (Sykes 2005, 2006a). It is, therefore, little wonder that hunting has come to be a character trait that is so frequently attributed to them.

Medieval Period (c. AD 1150–1550)

By the end of the twelfth century the elite of Western Europe were increasingly engaging with wild animals in multifarious ways: by maintaining them alive within private parks, by hunting them in a ritualized fashion with distinct hunting paraphernalia, by using their body parts for personal adornment (e.g. furs and feathers) or architectural display (e.g. antlers as hunting trophies), and also by employing their images as heraldic devices. Pluskowski (2007) has shown how the courtly behaviour and material culture of the time created a distinct but commonly understood seigneurial identity that spanned Europe in the later medieval period.

Figure 3.1 shows that wild animals are exceptionally well represented in archaeological assemblages of this date, reaching their highest frequency since the Mesolithic during the mid-fourteenth to mid-sixteenth century. However, it is evident from Table 3.1 that fallow deer are largely responsible for the increase as no other wild mammal taxa are well represented in this period. Utilization of the term 'wild' becomes debatable in the context, given that fallow deer were essentially farmed within the thousands of deer parks that were established between the twelfth and late fourteenth centuries (Liddiard 2003; Rackham 2000). Often these parks are viewed simply as larders, where game could be stored on-the-hoof ready for consumption at aristocratic feasts (Birrell 1992). Undoubtedly these animals did eventually make their way to the dining table but their significance and that of the parks from which they came, certainly went beyond food. As in the Roman period, these parks are a reflection of worldview, an expression of Christian dominance over the wilderness (Chapter 5).

Preece and Fraser (2000) have argued that the concept of 'dominion' as set out in Genesis 1, 26 represents a poor mistranslation into the English of the Hebrew term *rādâ*, which they argue does not equate to despotic subjugation of animals but rather to their stewardship. If it was originally intended that humans had a duty of care towards animals, it was a role that the medieval population fulfilled very badly. By the end of the sixteenth century, people had successfully hunted much of England's native fauna to the point of extinction. The lynx, bear, wolf, wild boar, beaver and many wild birds (e.g. cranes) were no longer present in the wild (O'Connor and Sykes 2010), red deer populations had fallen dramatically (Griffin 2007) and roe deer appear to have become extirpated across southern England, indicated by both zooarchaeological and genetic data (Baker 2011). With the absence of the *deor* that gave the term its name, the 'wilderness' no longer existed in any real sense, either physically or conceptually, in southern England.

Summary

The last eight millennia witnessed a transformation of attitudes to wild animals and nature, at least in southern England, from a situation when the 'wild' did not exist because it was *all* that existed, to a situation where it did not exist because humans had largely eradicated the elements that constituted the *wildeoren*. The transformation that took place between these two extremes was not, however, a linear evolution. In this chapter I have suggested that the dynamics and social significance of wild animal exploitation are connected to mortuary practice and wider attitudes to the natural world, which in turn correspond to the ebb and flow of human movement and/or long-distance contact.

It can be no coincidence that periods characterized by insular sedentism – notably the Iron Age and Early Anglo-Saxon periods – indicate an avoidance of, or reverence for, wild animals and the 'wilderness'. This contrasts dramatically with those periods characterized by movement and trade or diaspora – notably the Mesolithic, Roman and Norman Empires – which are also those characterized by hunting cultures. In these periods, mental geographies were broad and the wilderness was conceptualized as something to be engaged with, either as a statement of familiarity (e.g. for Mesolithic hunter-gatherers) or to emphasize power, authority and, particularly for the two Empires, control. However, even in periods where the prevailing worldview was one of human dominion over nature, wild animals – particularly those that were the focus of hunting – were not all perceived as inferior to people. The status attributed to hunting derived from the fact that the quarry and the landscape in which they dwelt were unpredictable and would not necessarily yield to an unworthy hunter. The fact that respectfulness was key is reflected by the votive offerings of wild animals, which are noted for all the cultures dating from the Mesolithic to the Early Anglo-Saxon periods.

Remarkable continuity is apparent in the cross-cultural associations between hunting, masculinity and warfare. The relationship becomes evident for the first time during the Neolithic and Bronze Age but the same concepts are then found recycled in the Roman, Saxon and medieval periods. By the mid-twelfth century hunting was the pop culture of the time, which gradually diminished populations of native quarry. These animals were replaced with new semi-domestic exotic game species such as fallow deer, which were maintained within equally tame parks, landscapes that became the target of so-called 'hunting' in the post-medieval period.

Changes in animals bring changes to the landscape, environment and daily practice, and the capacity of exotic animals to alter human society is a subject that will be examined in greater details in the next chapter, as we consider animal diasporas and cultural change.

Chapter 4

ANIMAL DIASPORA AND CULTURE CHANGE

With the rise of global trade, animals have found it increasingly easy to move beyond their natural ranges and, since the early 1800s, the number of 'alien fauna' – that is creatures existing outside their native range – has risen sharply with hundreds of exotic vertebrates and invertebrates becoming established worldwide (Genovesi *et al.* 2009; Hulme 2007). Research into these species has grown in parallel and there is now an expansive literature concerned primarily with the ecological and economic impact of alien biota (e.g. Mooney and Hobbs 2000). Reading these texts, one would be forgiven for assuming that the human-assisted diffusion of species is a modern phenomenon but natural historians have long demonstrated that people have been moving plants and animals, either purposefully or inadvertently, for millennia (e.g. Grayson 2001; Lever 2009; Rackham 2000; Yalden 1999). Social scientists are increasingly coming to realize that biodiversity and cultural diversity are intimately linked (Rotherham and Lambert 2011) and it is, therefore, possible to study animal biogeography as a proxy measure of human history, population movements and trade (Gardeisen 2002; Matisoo-Smith and Robins 2004; Mondini *et al.* 2004; Searle *et al.* 2009).

Research is being advanced immeasurably by increasingly sophisticated scientific methods – notably ancient DNA, isotope analysis and geometric morphometrics – that are providing ever tighter resolution on the timing and circumstances surrounding the movement of animal populations (e.g. Cucchi *et al.* 2005; Larson *et al.* 2012). These advances are impressive, and scientific analyses of past animal introductions should be encouraged as they likely hold the key to finding solutions to modern threats to biodiversity. However, archaeological studies of animal biogeography are often concerned more with identification and dating of exotic animals rather than investigating their meaning and influence. In many respects this is understandable as we cannot hope to appreciate the more complex issues of 'why' an animal was introduced without first knowing 'when' it was brought or by 'whom'. Nevertheless, the frequent preoccupation with finding the earliest examples of an introduced species (e.g. Cucchi *et al.* 2005; Storey *et al.* 2007; Sykes *et al.* 2006; Thomas 2010) is to overlook the capacity of exotic animals to alter human behaviour, worldview and cultural landscapes.

As was discussed in the previous chapter, anthropologists such as Helms (1993: 7) have demonstrated that traditional societies frequently equate geographical distance with supernatural distance, perceiving things derived from remote realms as powerful, carrying associations with gods, ancestors or cultural heroes. If this is the case for exotic objects, it seems unlikely that exotic animals – living things which 'act back' in a way that artefacts do not – would have been viewed in neutral terms. Instead it seems likely that they would have been afforded special attention, treatment and significance; essentially their arrival would have influenced human behaviour.

In our global culture of twenty-four-hour news and wildlife documentaries it is perhaps difficult to envisage a situation where the arrival of new animal species could have impacted upon society. For us, the only unknown 'outer realm' is outer space, and this is perhaps the best analogy to highlight how exciting the arrival of new things from beyond the limits of the known world must have been. Imagine if an extraterrestrial landed on this planet: such an event would surely transform world culture, requiring existing ideologies and cosmologies to be reconfigured in the light of new knowledge and technology. There is every reason to suspect that the alien beings would rapidly be incorporated into religious beliefs, as is already seen to some extent with the Church of Scientology (Urban 2011).

One does not need to look very far back in human history to find such scenarios being acted out, particularly during periods of European colonialism. Legend recounts that the first European explorers to land on the shores of the Americas were considered to be gods by the natives, who perceived the horse-riding Spaniards to be centaur-like hybrids (Restall 2003: 112). Similarly, the arrival of the European fur trade in Northwest America is said to have altered native cosmologies – whereby European goods (e.g. woollen blankets) took on ritual and ideological significance, even replacing the traditional animal skins in ceremonial activities (Helms 1993: 156). Certainly DeJohn Anderson (2004) has shown how the domestic animals brought to America by European settlers fundamentally transformed native lives and lifestyles. Similarly, although Australian Aborigines enjoyed stealing and eating the cattle and sheep brought by European settlers, they had high levels of respect and awe for the newly introduced dogs and horses, which were incorporated into, and transformed, their daily practice (Reynolds 1981: 51–3).

In this chapter I will review the evidence for animal introductions to Britain, whose island status renders it an excellent laboratory for investigating animal diasporas because we can be confident that the majority of introduced species were brought by people and therefore are relevant to the human story. Through a series of case-studies – specifically domestic livestock (cattle, sheep and pigs), the horse (*Equus caballus*), the chicken (*Gallus gallus*), the brown hare (*Lepus europaeus*) and fallow deer (*Dama dama*) – I aim to demonstrate how the arrival of exotic animals has the capacity to transform human lives and cultures.

Britain: A Case Study in Animal Diaspora and Culture Change

Eight thousand years ago, when the sea cut Britain off from the rest of the Continent, the island's fauna was very different. As was seen in the previous chapter, many animals that are now extinct were abundant whilst those familiar to us today were not present. New species were introduced through a variety of mechanisms: some brought purposefully by migrating human groups, others sent as royal gifts from far off lands, whilst several arrived as stowaways. Their stories provide a fascinating chronicle of human–animal relations, which are now beginning to be understood in increasing detail (Yalden 1999; O'Connor and Sykes 2010). For instance, whilst it is often suggested that new fauna were introduced by 'invading peoples' (the Romans and Normans being the most commonly invoked) it is now evident that the situation was rather more complex with many species – e.g. chickens, domestic cats, donkeys, house mice – being introduced during the (Late) Iron Age, when Britain was not under conquest.

Figure 4.1 summarizes current knowledge about the arrival dates of the main imported fauna. This reveals that the most significant episodes of species introduction occurred when Britain was well connected to Continental Europe: notably during the Neolithic, Bronze Age, Late Iron Age/Roman and medieval periods. Conversely, the periods during which Britain was more insular – the Early and Middle Iron Age and Anglo-Saxon period – show far less evidence for the introduction of new species. It is interesting to note that these diachronic patterns are reminiscent of those highlighted in Chapter 3, particularly those that show the rise and fall of wild animal exploitation (Figure 3.1) and the shifts in stable isotope analysis (Figure 3.2).

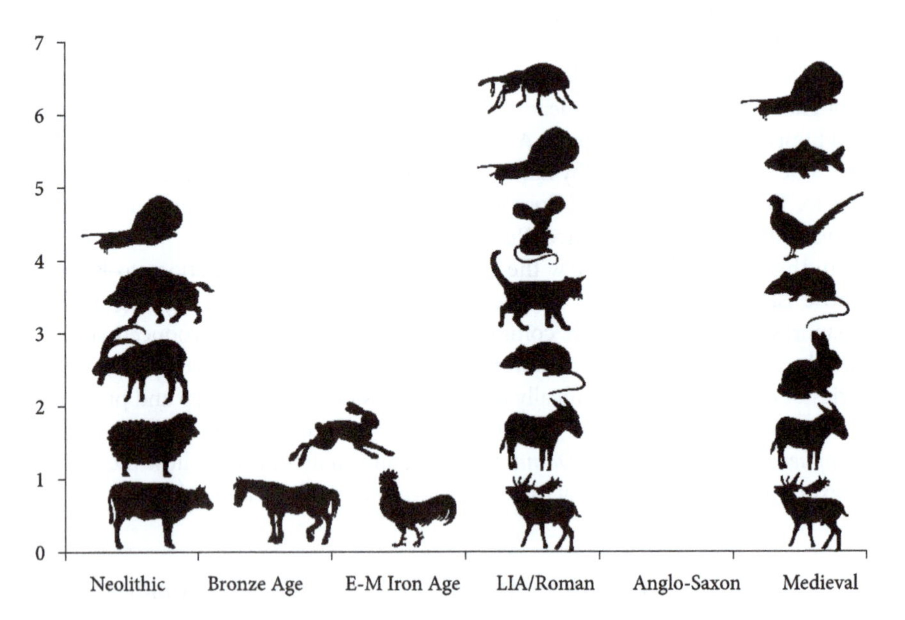

Figure 4.1 Diachronic variations in the number and type of species introduced to Britain.
Source: O'Connor and Sykes (2010).

The archaeobotanical record also shows very similar patterns in the representation of wild and exotic plants, recent research highlighting that Late Iron Age/Roman and medieval assemblages have far higher frequencies of imports – e.g. figs, olives, grapes, lentils, mulberry, pine nuts and exotic spices – than is seen in either Early/Middle Iron Age or Anglo-Saxon deposits (e.g. Livarda 2011; van der Veen 2008; van der Veen *et al.* 2008). The same is true for fruit trees, such as domestic apple and pear, which apparently became established in Britain for the first time during the Roman period (van der Veen *et al.* 2008, 33). According to van der Veen *et al.* (2008) these new plant species represent economic and dietary innovations with important implications for human nutrition and the way in which social and cultural identity were negotiated. Similar arguments have been proposed for the introduction of exotic animals (e.g. Sykes *et al.* 2006). Yet this concentration on exotics as 'products', 'nutrition' and 'symbols' is to overlook the inter-relationships that preceded these outcomes. As was set out in Chapter 1 (page 12), it is important to consider the life of animals and how they influenced human daily practice. This point is particularly relevant for exotic creatures, which were seldom introduced simply to enhance human diet (O'Connor and Sykes 2010).

Neolithic Livestock

The first domesticate livestock introduced to Britain – cattle, sheep, goats and pigs – played a far more fundamental role in human society than the mere provision of dietary protein. As was seen in Chapter 2, the Mesolithic-Neolithic transition in England was accompanied by a dramatic shift away from hunting and towards animal husbandry, in particular cattle husbandry (see Figure 3.1). The rapidity of this transformation cannot be explained simply by a desire to eat beef and mutton rather than venison. It becomes easier to comprehend, however, once it is accepted that these exotic animals are likely to have been viewed as very special indeed. Not only were they new and exciting – sheep and goats having no native wild progenitors in Britain – it is clear from their demographic profiles (which show an abundance of old females and juvenile males, Serjeantson 2011) and from lipid analysis of Neolithic ceramics (Copley *et al.* 2003) that, from the point of their introduction, cattle, and probably sheep and goats, were used for dairy production. Dairying may appear to equate directly with food but, within many traditional societies, dairy products are important components of magic and medicine (Crate 2008). These associations can be traced back through time; for instance, Anglo-Saxon Leechdoms (Cameron 1993: 8) frequently mention the use of milk, cheese and butter to cure a variety of ailments, and the Roman author Pliny the Elder provides extensive advice about the medicinal use of dairy products (*Natural History* XXXIII; trans Jones 1963: 123–31). It is inconceivable that the same was not true in the Neolithic, especially when the skills, technologies (e.g. pottery) and close human–animal relationship involved in dairying, as well as the novel liquid of milk itself, would have been unlike anything experienced in the Mesolithic. As Thomas (2004: 120) has suggested, all these elements – the animals and their associated products and paraphernalia – were probably understood as supernatural and magical, conferring new forms of power and status upon those in their possession. With this in mind, and given the traditional association between dairying, women and goddesses (Chapter 2) it is tempting to suggest that the introduction of domestic cattle and sheep/goats may have redefined gender relations in favour of women. Such a scenario may have been particularly marked in contrast to Mesolithic lifestyle, which presumably placed more emphasis on the traditionally male activity of hunting (Chapter 3). If domestication led to the empowerment of women, it would have given approximately 50 per cent of the human population great incentive for its speedy adoption.

Under such circumstances, domestic animals would have been deemed too valuable, both in terms of maintaining herd viability but also as active members of society, to be eaten. This accords well with the behaviour of many modern pastoral groups. Amongst the Suri and Me'en of Africa, for instance, domestic animals are rarely killed and usually only for particular rites of passage – birth, marriage, death – for divination, or to facilitate social intercourse and conflict resolution (Abbink 1993: 709; 2003: 348). If a similar situation occurred in Neolithic England, it could explain why the majority of animal bone assemblages of this date are restricted to ceremonial contexts – monuments and funerary structures – suggesting that the slaughter and consumption of domesticates took place only on special occasions of communal gatherings and feasting. Evidence to support this idea has recently been provided by strontium isotope studies of cattle teeth. Viner *et al.* (2010) examined teeth from thirteen individuals recovered from the Neolithic site of Durrington Walls (Wiltshire) and, of these, only two provided chemical signatures consistent with the local chalkland geology. The rest of the cattle appear to have been raised on very different geology, potentially coming from more than 100 km away. This suggests that cattle, and by proxy people, were occasionally mobile, travelling great distances to engage in large-scale communal feasting at particular ceremonial centres, in this case Durrington Walls. The act of converging on a site to share the meat from animals that were likely well known and highly valued, would have created community and encouraged social cohesion (Chapter 7). Furthermore, by depositing the remains of animals within enclosures and ditches, forms of public architecture that were not part of the Mesolithic landscape, the new relationships between humans and animals would have been inscribed into, and monumentalized within, the physical world (Serjeantson 2011: 53).

It is well known that Neolithic cattle remains were frequently deposited in ways that are indicative of ritual behaviour, with head and foot burials being particularly common in the ditches of causeway enclosures, pits and accompanying human burials (for a review see Serjeantson 2011). Morris (2008a) has surveyed the representation of such deposits, which he terms Associated Bone Groups (ABGs) and it can be seen that cattle ABGs reach their highest frequency (52 per cent) during the Neolithic period (Figure 4.2a). This is perhaps unsurprising since cattle are the best represented species in assemblages of this date (Figure 4.2b); however, Morris's work highlights some interesting disparities in ABG and overall assemblage representation for other animals and periods. One of the most notable of these relates to the horse.

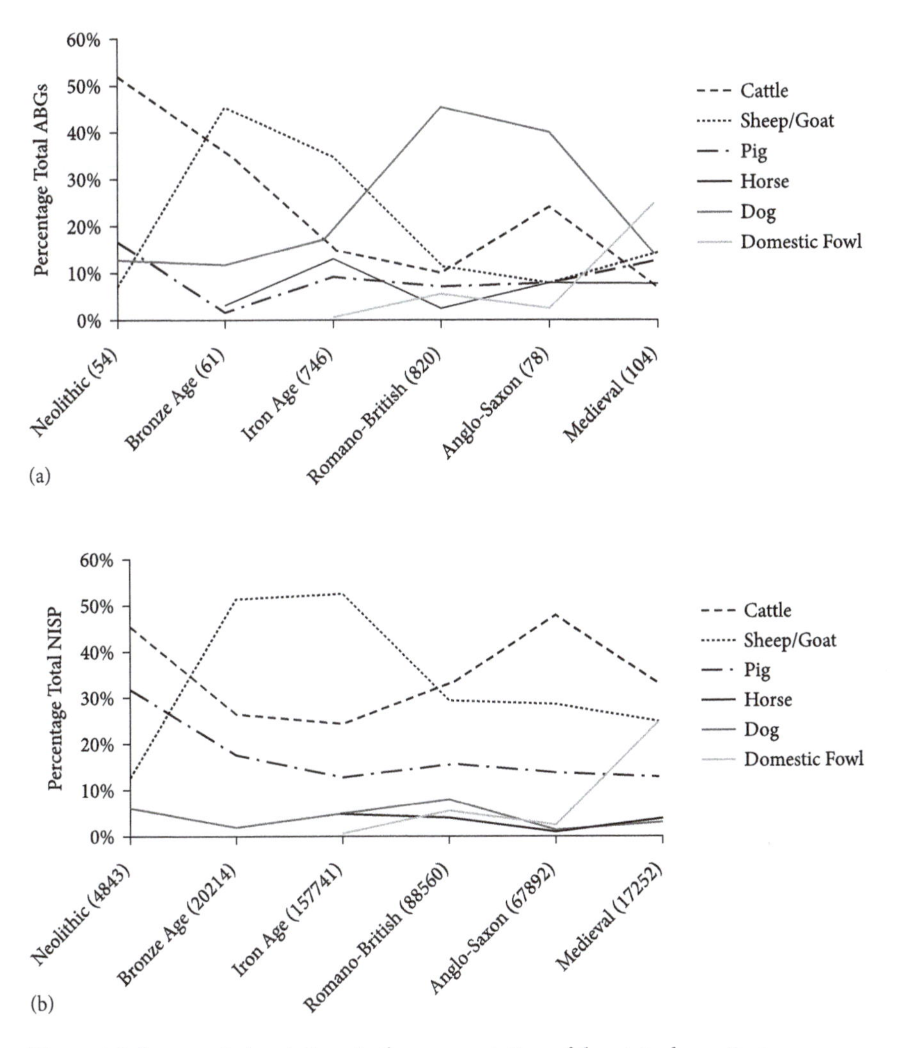

Figure 4.2 Inter-period variations in the representation of the main domesticates as percentage of (a) Total Associated Bone Groups and (b) Total NISP.

Source: Morris (2008a).

Horses of the Bronze Age and Iron Age

Horses are native to Britain but current evidence suggests that they became locally extinct at some point during the Mesolithic period: Bendrey *et al.* (2013) have demonstrated that there is no firm evidence that horses endured into the Neolithic period. Instead it would appear that they were reintroduced in small numbers during the Bronze Age, their populations becoming well established only towards the end of the Bronze Age and into the Iron Age. Figure 4.2b supports this suggestion, showing that horse remains are present in very low frequencies (*c.* 1 per cent) in Bronze Age assemblages but are more abundant in the Iron Age, although still making up a relatively low percentage (*c.* 5 per cent) of zooarchaeological assemblages as a whole. This situation contrasts with the ABG data (Figure 4.2a), which shows that horses are comparatively well represented in the Bronze Age (*c.* 4 per cent) and make up almost 14 per cent of the ABGs dated to the Iron Age. This disparity between the two datasets reveals that horses were afforded special treatment in death, and the fact that their representation as ABGs is particularly high in the period that follows their reintroduction to Britain indicates that this treatment was likely linked to their exotic status.

Ageing data support the idea that horses were highly valued. Overall, the age profiles of Bronze Age and Iron Age horses are skewed towards more mature animals and there is a marked absence of juveniles. Traditionally this has been interpreted as evidence that horse breeding was not controlled by people; that the animals were free range and were rounded up only periodically so that the best individuals could be selected for training (Harcourt 1979). Such a situation would explain the low frequencies of juvenile horses on Prehistoric sites, as the remains of infant fatalities would not be expected at settlement locations; however, Bendrey (2010: 15) has argued convincingly that the low number of juvenile animals could equally reflect a situation where skill and care in horse breeding saw the vast majority of animals surviving well into adulthood.

If viewed in this way, the ageing data suggest that human–horse relationships were very close in Prehistoric Britain, something that is also indicated by the frequent comingling of human and horse remains in Iron Age contexts, the most extensive survey of this widespread practice being provided by Hill (1995: 54). The mixing of human and horse is equally reflected by Iron Age iconography, Aldhouse-Green (2004: 160–4) and Creighton (1995, 2000: 45) providing a plethora of examples of Iron Age figurines and coinage that depict human-headed horses. The human–horse hybrid is a theme common to the mythology of most geographical regions where horses were introduced – from India, through the Near East to the Mediterranean and northern Europe. For instance the Vedic *Rigveda* describes the birth of the Aśvins, the divine horse-headed twin brothers, who were born to Saranya the Hindu cloud goddess and were responsible for the sunrise and sunset. The myth is closely related to the Ašvieniai, the twin horse brothers of Baltic mythology, who were also responsible for pulling the sun through the sky on a chariot. It also resonates with the sun horses of the Greeks, with the Greek Diskourori (twin sons of Zeus), Castor and Pollox, as well as the Anglo-Saxon brothers Hengist and Horsa, all of whom were closely associated with horses (West 2007: 186–91).

Perhaps the most famous of the human–horse hybrids is the centaur, which according to Greek legend was born when Ixion, king of the Thessalian Lapithe, slept with a cloud that Zeus had shaped to resemble the goddess Hera (Padgett 2004: 14). The origin of the Greek centaur myth has long been debated. Iconographic representations of centaurs become increasingly common on Greek vases from the seventh century BC onwards; however, there are earlier examples. The Lefkandi

Centaur, a tenth-century BC terracotta figure recovered from Euboia, is thought to represent the mythical centaur Cheiron (Padgett 2004: 14). If this identification is correct it would indicate that stories of centaurs were in oral circulation long before they were first recorded. More recently Shear (2002) has been able to push the origins of the centaur myth back to the Bronze Age, based on the discovery of Mycenaean terracotta centaurs from Ugarit, which have been dated to the Late Helladic period. Given that horses were introduced to Greece in the Early/Middle Helladic period it seems probable that it was their arrival that prompted the earliest stories of centaurs (Shear 2002: 152).

The possibility that the centaur myth was introduced to Greece fully formed from the East is indicated not only by a Middle Assyrian seal dated to the thirteenth century BC, which clearly depicts a centaur, but also by the similarity of the centaur myth to that of the horse-headed Aśvins of Vedic literature, already discussed, whose conception was equally linked to clouds (West 2007: 192). Both stories are also reminiscent of Siberian origin myths: the Sakha of Siberia believe the horse to have been the first creation made by the highest god of the sky pantheon, from which came creatures that were half-man half-horse and from there the human race was born (Crate 2008: 117). Given the widespread and similar nature of horse-related mythology it is not inconceivable that the horse–human iconography and combined human–horse burials of Prehistoric Britain reflect similar traditions.

Creighton (1995, 2000: 2) has suggested that the iconographic and archaeological evidence for British Iron Age horses reflect rites of power legitimization, whereby leaders sought to validate their authority through association with an animal that was clearly viewed as semi-divine. This is nowhere better exemplified than by the kingship ritual of the Aśvamedha, which is recounted in the Vedic literature but finds compelling, and often cited, comparanda in the Roman 'October *Equus*' and the Irish kingship inauguration 'Feast of Tara', the latter detailed in early medieval literature (Watkins 1995: 267). All three rites involve sacrifice of a prize horse, with the Indian and Irish kingship rituals having the added bonus of requiring the would-be king, or his wife, to copulate with the animal, either symbolically or actually. The animal was then cooked and eaten by the king. These acts are believed to represent the ruler's union with and embodiment of the forces of nature (represented by the horse), thereby legitimizing the king's supremacy over the cosmic order (Creighton 2000: 22).

The possibility that such rituals may have been introduced to Prehistoric Britain as part of a new horse-culture is far from certain but the idea is given credence by finds of first century BC coins, admittedly from central Gaul, that carry the inscription *IIIPOMIIDVOS*. This translates as 'Epomeduos', a term that is the Gaulish equivalent of 'Aśvamedha', suggesting that the rite was known in northern Europe during the Iron Age. The fact that it was referenced on coins – an important medium through which power was communicated – reinforces the association between horses, leadership and the supernatural (Creighton 2000).

Linguistics provide perhaps the strongest evidence for a common horse culture: philologists have been able to demonstrate that the Anatolian, Tocharion,

Indo-Iranian, Italic, Celtic, Germanic, Celtic and Greek words for horse all derive from the Proto-Indo-European term **ekwos* (Winter 1997: 435). Wider scholarship concerning horse-related mythology, ritual and linguistics have all highlighted startling similarities across geography, which suggests that the diffusion of the horse was accompanied by specific cultural elements that were incorporated into the practices and belief systems of the adoptive societies (Creighton 2000; O'Flaherty 1980; Skjaervø 2008; Watkins 1995; West 2007). In this way, the arrival of the horse to Britain must have been truly transformative, bringing new mechanisms for the negotiation and legitimization of power, and this is to say nothing of the impact that horses would have had on human warfare, movement and perceptions of landscapes (Bendrey 2010). As such, the horse is an excellent example of an animal having the capacity to genuinely change human life and worldview.

The ability of animals to impact on human daily practice is something that I understand well, having introduced a dozen free-range hens into my garden (and life). Their arrival has transformed not only my world but also my research, which is now increasingly focusing upon cultural perceptions of the chicken.

Chickens in the Iron Age and Roman Period

Given the modern global distribution and significance of the chicken, astonishingly little is known about the exact timing, circumstances and rationale of their introduction to Britain and a detailed study of this species' spread across Europe is long overdue. What is clear is that, contrary to popular belief, the Romans were not responsible for the introduction of the chicken to Britain, where they were certainly established during the Iron Age (Poole 2010a; Sykes 2012). Based on modern British farming and consumption practices, logic would suggest that chickens were introduced for meat and eggs but, whilst this interpretation is economically rational, it seems at odds with all the evidence. From the point of their introduction, the contexts in which chickens are recovered suggest that chickens were not eaten but rather held a special status: Figures 4.2a and 4.2b show that chickens are better represented in ABGs than in general assemblages. Furthermore, Morris (2008b: 87) has identified a link between deposits of complete/partial domestic fowl skeletons and Late Iron Age human graves. Further analyses (Doherty 2013; Sykes 2012) indicate that these funerary practices are gendered with men and women being interred with cockerels and hens respectively. Domestic fowl continued to be linked with religious practices into the Roman period, when they were closely associated with the cult of Mithras and the god Mercury (Crummy 2007): large quantities of cockerel remains were recovered from the *Mithraea* at Walbrook in London and Hadrian's Wall, Northumberland, as well as the shrine to Mercury at Uley, Gloucestershire (King 2005).

Such a situation should be unsurprising given that everywhere else that chickens have been introduced, taboos, either cultural or legal, have existed against their consumption (Chapter 8). For instance, in West Africa, where domestic fowl became widespread only after AD 1000, their ritual significance and role in magic outstripped their dietary value (MacDonald and Blend 2000: 496). Similarly when domestic fowl were introduced to the Americas (the date of this still being a matter of contention – see Storey *et al.* 2007; Gongora *et al.* 2008; Fitzpatrick and Callaghan 2009) it seems that they were rarely used for meat and eggs but more often for feathers and as pets (Seligmann 1987), with cockfighting being well established by the sixteenth century (Cobb 2003: 73).

The association between the spread of domestic fowl and cockfighting is indicated by a variety of evidence strands. For instance, Liu *et al.*'s (2006) phylogeographical study of chickens identified a clade of gamebirds whose distribution across Asia correlates closely with the geographical distribution of cockfighting societies. Ancient literature and iconography attest to the importance of cockfighting in India, where domestic fowl were present, but not obviously eaten, from 2500 BC (Somvabshi 2006: 141; Zeuner 1963). Domestic fowl begin to appear regularly in the archaeological, iconographic and historical records of western Asia and the Mediterranean during the first millennium BC. Often their representation is specifically in the context of cockfighting, such as in Israel, Palestine and Egypt where cockerels and cockfights are depicted on seventh-century BC seals and ceramics (MacDonald and Blend 2000: 497). Here all the evidence points to domestic fowl being used exclusively as fighting birds, rather as a source of food (Crawford 2003: 12). Indisputable evidence for cockfighting comes from ancient Greece where its cultural importance as a symbol of warfare and (homo) sexuality is attested by both written and pictorial sources (Shelton 2007: 102–6).

It is widely assumed that the Romans adopted the Greeks' passion for cockfighting, spreading the practice across their Empire (e.g. Serjeantson 2009: 326–30) but this assumption has been challenged by Morgan (1975: 121) based on the 'absolutely deafening silence of so many Roman writers'. True, the documentary and iconographic evidence for cockfighting is not as abundant as for ancient Greece but it is there – Figure 4.3 is one of many Roman depictions of cockfighting. Although Morgan's (1975) argument – that such illustrations may have had more complex meanings than simply reflecting daily behaviour – is highly valid, the archaeological evidence provides a better indication of demotic practice and it would appear to support the case that cockfighting was a significant activity in the Roman Empire. For instance, zooarchaeological assemblages from the Roman castellum at Velsen in the Netherlands produced an exceptionally high percentage of cockerels to hens, as did the assemblages from the Romano-British towns of York (Yorkshire), Dorchester (Dorset) and Silchester (Hampshire). Given that artificial cock-spurs were recovered from Silchester (Serjeantson 2009: 329) it seems likely that cockfighting did take place within urban environments. Other artificial cock-spurs have been recovered from sites in Cornwall (Scott 1957: 53) and at Baldock in Hertfordshire an iron cockspur was found still attached to the cock's leg (Hingley 2006: 231). The possibility that the origin of cockfighting in Britain dates back to the Iron Age introduction of the cockerel is indicated not only by the fact that the earliest examples of chickens in Britain contain a fairly high proportion of adult cockerels but also by Caesar's statement in his *The Gallic*

Figure 4.3 Late-Antique mosaic depicting cockfighting from Baptistery in Butrint, Albania.

Source: Martin Smith (Butrint Foundation).

War (V.12 trans. Handford 1982) that the Britons 'do not regard it lawful to eat the hare, and the cock, and the goose; they, however, breed them for amusement and pleasure'. Conceivably the 'amusement' may have been cockfighting.

Given that most strands of evidence intimate that cockfighting was a central impetus for the domestication and spread of chickens it is interesting that there has been little investigation about how the arrival of a cockfighting culture may have impacted upon the societies that adopted it. The lack of analysis is all the more surprising considering the large amount of anthropological work on the subject, to which we now turn.

Anthropological studies have shown repeatedly that wherever cockfighting is practised it is usually central to the construction of cultural identity or gender alignment, particularly the definition of masculinity (Cobb 2003; Dundes 1994; Hicks 2006/7). Indeed, Marvin (1984) has argued that cockerels are symbols of masculinity as they embody all the qualities of the 'truly male'. Cobb (2003: 77) states that cockfighting 'wears misogyny on its sleeve' and it is not surprising that, today, cockfighting tends to be a male-only sport. In general terms it is more common in patriarchal societies, where rates of female-directed violence or 'intimate terrorism' are comparatively high (Johnson 2006: 1015). In some cases, domestic violence has been linked directly with cockfighting (McCulloch and Stancich 1998). These findings concur with the growing body of evidence (e.g. Mullin 1999) that suggests the way that people treat animals is a reflection of, or at least provides an insight into, how they treat one another; Henry (2004: 423) specifically mentions cockfighting in his article on the relationship between animal cruelty and human delinquency.

The idea that humans and animals become mutually socialized through their interactions is interesting with regard to chickens and it was certainly the conclusion of Geertz (1994), whose pioneering study of cockfighting in Balinese society highlighted the close identification between men and their birds, the two being almost indivisible. This permeability between bird and keeper is reinforced by Guggenheim's (1994) work on cockfighting in the Philippines where it is believed that both cockerels and keepers should abstain from sex prior to fights to preserve vital energies but that sex is heartily recommended for all parties afterwards. Not all anthropological works have found cocks to be the embodiments of their owners. Notably, Marvin's (1984) study of cockfighting in Andalusia concluded that, although cockfighting is central to the construction of masculine identity and that key qualities of cockerels – their assertiveness, courage and tenacity – are valued as male attributes, there is no permeability between man and cockerel. Instead Marvin (1984: 60) suggests that cockerels and cockfighting are more important for displacing human aggression onto animals; that human society is less violent as a result. A similar interpretation could be drawn for Hicks' (2006/7) analysis of East Timor where cockfighting to-the-death has replaced the lethally violent tradition of headhunting with its metaphorical links to masculinity, fertility and ancestral validation, which are now also conferred upon cockfighting.

Having briefly outlined some of the anthropological literature and sociological theories surrounding cockfighting, as well as human–animal studies more generally, my question is this: is it possible that the introduction and establishment of domestic fowl, an exotic species whose behaviour is unlike any of Europe's native fauna, could have altered human behaviour and ideology? If, as the anthropological and archaeological evidence seem to indicate, early imports of domestic fowl were viewed as 'special', their arrival and establishment in Britain could have been socially influential. It seems feasible that a male-centric cockfighting culture could have arisen, potentially bringing an associated increase in female-directed violence, particularly if the sexually dominant behaviour of the cockerels was adopted by those who interacted with them. On the other hand, it might also be expected that rates of interpersonal violence would fall as cockfighting became a mechanism for conflict resolution, as has been suggested for Andalusia (Marvin 1984) and East Timor (Hicks 2006/7).

The connection between cockfighting and headhunting is potentially interesting within the context of Iron Age Europe, where a headhunting or, in the case of Britain, skull-curation culture is now well attested, especially for the earlier parts of the period (Armit 2011: 512). Armit's (2011: 512) description of headhunting as 'associated with cosmological ideas of rebirth and fertility' is very close to that which Hicks (2006/7: 16) gives for both headhunting and cockfighting in East Timor, which he states represent 'a complex of ideas involving life, fertility, death, violence, blood and ancestral authority'. For Iron Age Britain, Redfern (2008) has shown that skull fragments from adult male individuals, many of whom appear to have suffered blunt-force cranial injuries, are incorporated into structured deposits in Early and Middle Iron Age Dorset, but that this phenomenon declines into the Later Iron Age, when the majority of individuals were buried as articulated inhumations. As has already been shown above, some of these Late Iron Age graves contain domestic fowl and it is interesting that there is an inverse correlation between their representation and the evidence for skull curation.

More convincing evidence for an Iron Age to Roman decline in interpersonal violence is provided by Redfern's (2006) extensive research on skeletal collections from Iron Age and Roman Dorset, the only region of Britain where there is pre- to post-Conquest continuity in burial practice. Her study of trauma rates, in particular those for sharp-force weapon injuries, demonstrated a statistically significant drop in prevalence from the Iron Age to Roman period in both men and women, suggesting that interpersonal conflict did indeed decline during the Roman period. Traditional interpretations would suggest that this was due to *Pax Romana*, the peace of Roman rule in the Empire, but a more faunal-based interpretation could be that the emergence of a cockfighting culture allowed human conflict to be displaced onto animals. I am not suggesting that domestic fowl were the 'peace-makers' of Iron Age and Roman Europe but rather that cockfighting reflected wider social shifts associated with Rome where, from the Middle Republic period (264–133 BC) onwards, animals and animal-like humans (e.g. prisoners of war, slaves and criminals) were increasingly utilized in violent displays within the arena – perhaps a more civilized version of headhunting (Hughes 2003; Kyle 2003).

As with the cockfighting-headhunting/skull curation theory, my hypothesis that female-directed domestic violence might have increased with the introduction of cockfighting is consistent with traditional interpretations of the social change that accompanied the Iron Age/Roman transition in Britain: for instance that the status of women gradually declined relative to men during the Roman period (Redfern and DeWitte 2011: 279). Whilst Redfern and DeWitte actually argue against the idea that the status of women was adversely affected during the period of Roman colonization, Redfern's (2006) research indicates that the prevalence of ante-mortem nasal fractures in females was higher in the Roman than Iron Age assemblages. Comparison with contemporary clinical data indicates that these kinds of fractures are associated with domestic violence (Redfern 2006: 379) and, although the dataset for Iron Age and Roman Britain is currently exceptionally limited, recent research (Doherty 2013) has strengthened the connection between cockfighting and female-directed violence in Roman Britain. As Serjeantson (2009: 330) has pointed out, cockfighting has always conflated sex and violence and it seems possible that this rise in violence towards women was encouraged by the arrival of a new masculine cockfighting culture.

So far, my discussion has centred on domestic animals; however, with my last two case-studies I wish to examine the introduction of species that are more frequently classified as wild. The first is the brown hare, *Lepus europaeus*, which is, to some extent, affiliated with domestic fowl, particularly in the form of the Easter egg and the Easter hare (now bunny), an association that is discussed further below.

The Brown Hare

Britain is home to two species of hare: the mountain or blue hare, *Lepus timidus*, and the brown hare. The mountain hare is certainly native to Britain but DNA studies (Stamatis *et al.* 2009) indicate the brown hare to be a more recent incomer:

its genetic variability is very low, suggestive of a small founder population. It is possible that the brown hare arrived in Britain naturally via the land-bridge that existed in the early Holocene; however, there are reasons to believe that it was brought by people at a later date.

Zooarchaeological evidence for the representation of hares is summarized in Table 3.1 but these data must be viewed with caution. First, this is because the results for the two hare species are conflated, as their remains are seldom separated by zooarchaeologists. This is perhaps not too problematic as the mountain hare is considered native only to the Scottish Highlands, so is unlikely to be widely represented in archaeological assemblages (Yalden 1999: 127; 2010: 194). The second caution relates to the meaning of the data: it must be remembered that while they reflect patterns of human exploitation, they need not indicate the presence/absence of the hare in the British landscape. Despite these caveats, it is clear that hares are very well represented in animal bone assemblages dating to the 'hunting cultures' of the later medieval and Roman periods, when hare remains are found in about 50 per cent of assemblages (Chapter 3). The Romans are most frequently cited as importers of brown hare and certainly there is evidence that Mediterranean practices of hare farming were introduced during the Roman period: the high levels of hares recovered from Fishbourne Palace (Sussex) and Whitehall Villa (Northamptonshire) have been interpreted as evidence that hares were being raised within local parks or *leporaria* (Sykes 2010a). However, use of hares for human sustenance is at odds with the idea that the Romans were responsible for the introduction of this species. As we have seen, exotic animals are seldom imported to provide protein for the human diet; indeed, their consumption is frequently tabooed. Caesar's comment that hares, along with chickens, were not eaten by the ancient Britons (see above page 86) confirms that hares were already established in Iron Age Britain and indicates that they had a significance beyond food, as might be expected for a recently arrived animal from the 'outer realm'.

The possibility that hares became established in Britain during the Iron Age is suggested by Table 3.1, which shows that their representation first became pronounced in this period, when they are represented in 18 to 39 per cent of assemblages. Prior to this, hare remains are rare, although their frequency does increase markedly in middle/late Bronze Age (from 4.7 to 11 per cent), which may indicate that they were first introduced in small numbers at this time, as Yalden (2010) has suggested. As was seen in Chapter 2, neither the Bronze Age nor Iron Age witnessed strong hunting cultures, quite the opposite, and therefore the high representation of hares, particularly in Iron Age assemblages, stands out as interesting.

On the basis of contextual evidence, it seems possible that rather than being hunted animals, these hares were prized exotica kept by people as pets, as Crummy (2013) has argued. Unlike the situation in the Roman period, when hare remains are most frequently recovered from domestic deposits indicative of food waste (see Figure 3.1), the majority of Iron Age hares consist of complete skeletons, or ABGs, that show no evidence of butchery marks (Crummy 2013). To be fair, hare ABGs are also found in Roman Britain: Morris (2008a: 152) notes eight examples from four different sites in Hampshire and Wiltshire, whilst King (2005) notes that hare bones have been recovered from Roman temples across southern Britain. The geographical range of hare ABGs corresponds neatly with the distribution of hare brooches, which Crummy (2013) has shown cluster chiefly in Hampshire with additional outcrops in Norfolk, Leicestershire and Lincolnshire. Zoomorphic plate brooches have been shown to reference specific deities (Crummy 2007) and, on this basis, Crummy (2013) has suggested that a hare-deity with origins in Celtic folk religion was worshipped in these regions. When united, the zooarchaeological and artefactual data provide reasonable support for the idea that the brown hare and an associated hare deity were present in Iron Age Britain. If this is the case, the next question to be solved is: from where were they introduced? The answer will probably not be forthcoming until a comprehensive programme of ancient DNA work is undertaken; however, modern hare genetics may provide some clues.

Stamatis *et al.* (2009) surveyed almost 1,000 modern hare specimens from thirty-three locations across Europe. They found that British hares are closely related to populations in northern Germany, which Stamatis *et al.* (2009) suggest is the source region for British hares. More specifically, in their investigation of hares from northern Britain, Stamatis *et al.* (2009) identified one haplotype that elsewhere is found only in one north German population, close to the Danish border. The suggestion that British brown hare populations may have been introduced from this region is interesting in the light of Shaw's (2011) study of the goddess Eostre, from which the name Easter derives and with whom hares have traditionally been associated, albeit without much evidence (Boyle 1973: 323).

There has been considerable debate about whether Eostre was a real pre-Christian deity or whether she was constructed and popularized by the Anglo-Saxon author, the Venerable Bede, whose *Reckoning of Time* provides the main evidence for the goddess's existence. In this text Bede lists the names of the English months together with their etymologies, explaining that April correspond to the English *Eosturmonath*, apparently so called 'after the goddess of theirs, named Eostre, in whose honour feasts were celebrated in that month' (Wallis 1999: 54). Shaw (2011) provides an extensive analysis of the evidence – linguistic, place/personal-name and epigraphic – for and against Eostre. He concludes that, in all probability, a cult of Eostre did exist in certain regions of Anglo-Saxon Britain, the place-name evidence suggesting Kent as a possible focus. Shaw also suggests that a similar cult was present in the Rhineland. This is based on the recovery of over 150 Romano-Germanic votive inscriptions made to a group of female deities named the *matronae Austriahenae*, which were found near Morken-Harff in north Germany and dated to between AD 150 and 250. The first element of name *Austriahenae* can be connected etymologically with Eostre, suggesting that the two share some parallels, and Derks (1998: 128) has argued that the *Austriahenae* developed from a pre-Roman ancestor cult. Shaw (2011: 61) warns against interpreting this as evidence for a pan-Germanic cult or using it to suggest that the Anglo-Saxon Eostre developed from the Roman (or pre-Roman) *Austriahenae*. However, given that the genetic data suggest that the brown hare (so commonly associated with Eostre) was introduced to Britain from the same region where we find early evidence for the

Austriahenae cult, it is exceptionally tempting to draw such an interpretation. I should say that both Nina Crummy and Phillip Shaw have recommended that I resist this temptation at all costs, and I suspect they are right. Nevertheless, I hope this case study has highlighted the severe knowledge gap surrounding the brown hare: research into this species would undoubtedly be worthwhile, with potential to transform traditional scholarship across a range of disciplines.

My next, and final, case study in this chapter is based on a more sure footing as the species in question – the fallow deer – has been the subject of my research for over a decade.

The Fallow Deer of the Roman and Norman Empires

Of all the World's deer, the distribution of the European fallow deer (*Dama dama dama*) has been most influenced by humans, who have transported the species from its Mediterranean glacial refuge across Europe, Africa, America and Australasia (Chapman and Chapman 1975). Zooarchaeological evidence suggests that fallow deer were introduced to the island of Rhodes as early as the sixth millennium BC (Masseti *et al.* 2006), with populations being established across the Balkans and the Aegean in the Bronze Age: fallow deer were incorporated into the Mycenaean and Minoan cultures and they are represented iconographically, mythologically and zooarchaeologically, especially within Greek votive deposits (Yannouli and Trantalidou 1999). The first documentary evidence for the management of fallow deer is provided by Roman authors, such as Columella and Varro, and it is to the Romans that the popularization and spread of the deer management concept should be attributed: *Dama* remains have been recovered from Roman period sites in Italy, Sicily, Portugal, Switzerland, France, the Netherlands and Britain (Sykes *et al.* 2011).

Roman fallow deer remains are, however, rather rare in northern Europe and there are often questions over their status, as there exists the possibility that they represent traded body parts rather than animals that lived and died in the local area (Madgwick *et al.* 2013). Certainly during the Roman period, and even in the Iron Age, there appears to have been a lively trade in fallow deer body parts, with skeletal elements being imported from southern to northern Europe. This is indicated by Table 4.1, which shows that whereas Roman assemblages from southern Europe contain all parts of the fallow deer skeleton, assemblages from northern Europe are dominated by shed antlers and metapodia – elements that appear to have been viewed as portable objects in their own right. Long distance trade in fallow deer body parts has recently been confirmed by isotope studies conducted on shed antlers and foot bones from several sites in Britain, which have yielded results that are inconsistent with UK deer but plot closely with the expected ranges for the Mediterranean (Madgwick *et al.* 2013). The rationale for this trade in antler and foot bones is discussed further in Chapter 6 and may be related to the classical belief that deer bones had protective and medicinal properties. However, the recovery of several north European specimens from votive deposits, e.g. the three foot bones from Augst in Switzerland (Schmid 1965) were found associated with a

Table 4.1 Skeletal representation data for all known *Dama* specimens from Iron Age Britain, Roman Northern Europe and Roman Southern Europe

Site	**Antler**	**Mandible**	**Scapula**	**Humerus**	**Radius**	**Metacarpal**	**Pelvis**	**Femur**	**Tibia**	**Astragalus**	**Calcaneum**	**Metatarsal**
Lydney, Gloucestershire, UK	1											
War Ditches, Cambridgeshire, UK	2											
Iron Age Britain Total	**3**											
Monkton, Isle of Thanet, Kent, UK	2		2		1	3						1
Fishbourne, Sussex, UK	3	2	3	1	1	10	1	1	1			1
Canterbury (St Georges), UK												1
Canterbury (Whitefriars), UK			1		1							
London (Salvation Army site), UK					1							
Cowdrey's Down, Hampshire, UK					1							
Scole-Dickleburgh, Norfolk, UK	1											
Dorchester-on-Thames, Oxfordshire, UK	1											
Wroxeter, Shropshire, UK									1			
St Albans (Park Street), Hertfordshire, UK	1											
Catsgore, Somerset, UK						1						
Valkenburg, Limburg, Netherlands												1
Rouen, Normandy, France										1		
Mauregard, Seine-et-Marne, France	1											

Chartres, Picardie, France	1											
Savy, Picardie, France	1											
Augst, Switzerland						2						1
Roman Northern Europe Total	***11***	***2***	***6***	***1***	***5***	***16***	***1***	***1***	***2***	***1***		***5***
São Pedro, Lisbon, Portugal									1			
Torre de Palma, Lisbon, Portugal				1					2	2		
Vilauba, Girona, Spain												1
S'Illot, Mallorca		1	1			1		1	1		1	1
Le Colonne, Tuscany, Italy	1											
Settefinistre, Tuscany, Italy	2	1			1		1					
San Potito-ovindoli, L'Aquila, Italy					1							
Monte Gelato, Lazio, Italy					2	1						
Girolamini, Naples, Italy	1											
Santa Maria La Nova, Naples, Italy					1							
Carminiello ai Mannesi, Naples, Italy					1							
Castanga, Sicily, Italy		4	2	1	5	6			10	5	4	2
Roman Southern Europe Total	***4***	***6***	***3***	***2***	***11***	***8***	***1***	***1***	***14***	***7***	***5***	***4***

temple to Diana, the goddess with whom fallow deer were intimately associated (Figure 3.3b; Reinken 1997; Yannouli and Trantalidou 1999). Indeed, it is becoming increasingly likely that it was the fallow deer's association with Diana, and her earlier incarnation as Artemis, that drove the not only the trade in body parts but also the spread and establishment of live fallow deer populations (Sykes in press).

It is noteworthy that Table 4.1 highlights two UK sites where the anatomical representation of the *Dama* assemblage does not conform to the usual antler-metapodia pattern: Fishbourne Roman Palace in Sussex, purportedly the residence of the Iron Age king *Togidubnus*, and Monkton, located on the Isle of Thanet in Kent. The fallow deer remains from both sites have been confirmed as Roman through radiocarbon dating and the application of other scientific methods indicate that all derive from animals that lived and died in Britain. In relation to Fishbourne, Sykes *et al.* (2006) identified early introductions to Fishbourne, with mandibles dated to 60AD±40 and 90AD±40. Strontium isotope analyses of the teeth provided convincing evidence that the earlier of the two mandibles was from a first generation introduction: the Sr ratio for the first molar demonstrated a non-local signature whereas the signature from the third molar indicated that the same individual was likely to have arrived in the Fishbourne area during sub-adulthood, where it then lived out the rest of its life (see Figure 4.4). Recent oxygen isotope analysis of the first molar suggests that the animal was raised in the central or western Mediterranean, and genetic analysis of both the Fishbourne and Monkton assemblages would appear to confirm this suggestion, the ancient DNA data mapping more closely with modern Italian deer than Anatolian populations (Sykes *et al.* 2011; Baker personal communication).

Explanations as to why the Romans went to such efforts to transport and maintain fallow deer can perhaps be found in the writings of classical authors. For instance, Columella, who was writing in AD 70 (approximately the same date that Fishbourne's fallow deer herd was being established), stated that:

> wild creatures, such as roe deer, chamois and also scimitar-horned oryx, *fallow deer* and wild boars sometimes serve to enhance the splendour and pleasure of their owners.
>
> —(*De Res Rustica* Book IX, I.1 trans. Forester and Heffner 1955, my emphasis)

This impetus for keeping wild beasts is echoed by several other classical authors (for instance Varro *Rerum Rusticarum* Book III, 13.2–3, trans. Hooper and Ash 1979). Certainly, the emparkment of fallow deer would have been a potent statement of socio-economic status because it married together three elements that have repeatedly, and cross-culturally, been incorporated into expressions of power and authority. The first of these is the 'wild' status of fallow deer. Although fallow deer are predisposed to taming – indeed the name *dama* is thought to have the same origin as the word 'tame' (Reinken 1997) – in the Roman mind they were perceived as coming from outside the domestic sphere (Starr 1992: 438). Their social meaning would have been defined further by the fact they were exotic – the

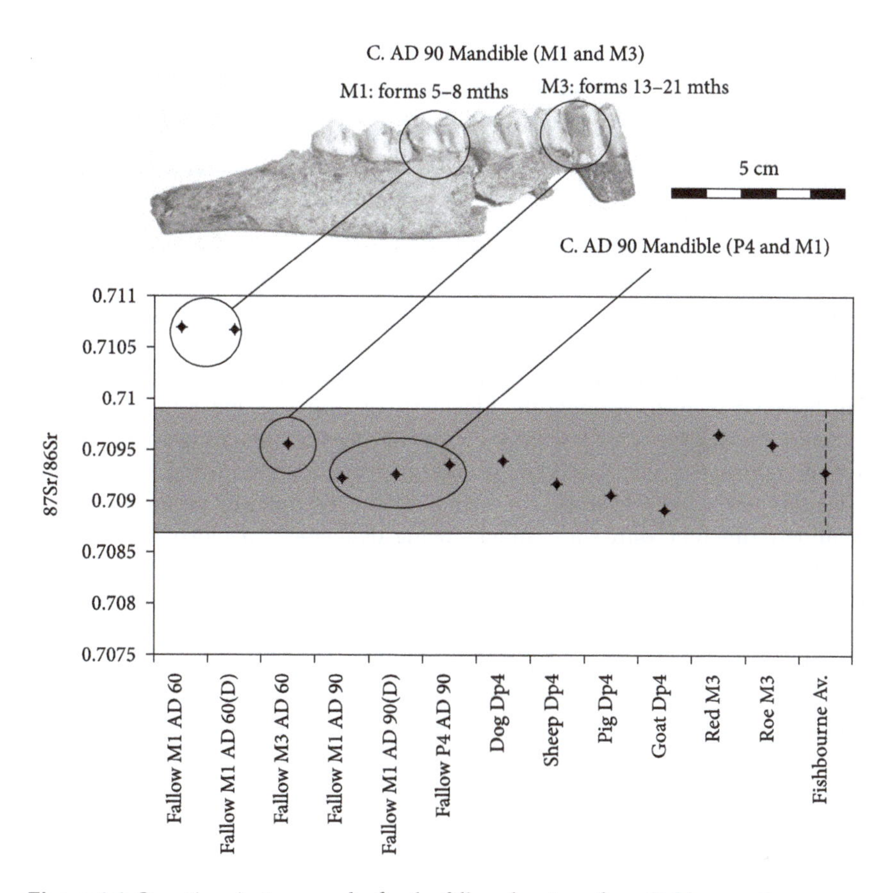

Figure 4.4 Strontium isotope results for the fallow deer jaws from Fishbourne Roman Palace. The shaded area represents the 'local' Sr signature for Fishbourne.

Source: Sykes.

second significant element: foreign commodities, which carry connotations of travel and suggest knowledge of distant lands, are frequently used to enhance rank and social standing (Goody 1982; Helms 1993). The third element to reinforce the social significance of fallow deer was their maintenance within a private space. Creation of a bounded park, an environment of pleasure and plenty separated physically from the outside world, would have been particularly divisive, communicating authority to those who were invited into the park but more particularly to those excluded from it.

Together, these three elements (wild, exotic and private) would have made fallow deer an icon of social position and it may be assumed that everything they represented was reproduced during their capture and consumption: the contextual evidence indicates that these animals were eventually eaten. But it is important to see them as representing more than just venison. Ageing data indicate that the majority of deer were kept for many years, some into old age. This would have allowed human–deer relationships to build and it may be that the fallow deer came to be viewed as pets, as is detailed in Virgil's *Aeneid* (Book VII, 483–6) which describes how a young girl, Silvia, kept a pet stag that was captured as a newborn (Starr 1992). Other historical sources mention how some estate owners choreographed the feeding of park animals to entertain guests as they dined *al fresco*: according to Varro (*Rerum Rusticarum* Book III.13.1 trans. Hooper and Ash 1979), emparked deer and boar were trained to assemble for food when a horn was blown, much to the delight of the audience.

At the most basic level the introduction of fallow deer and parks indicate that people perceived that they were *worth* the expense of their upkeep – this itself is a cultural stance. More importantly, however, the presence of animal parks indicates that people believed they had the *right* to enclose wild animals. This is a key point which would seem to separate the ideologies of Iron Age and Romano-British societies. As was seen in Chapter 3, ownership of wild animals appealed to the Roman mindset, which saw human control over nature as a sign of sanctity (Cartmill 1993: 52). However, the beliefs of the pre-Roman population were seemingly different with wild animals being deliberately avoided in Iron Age Britain. If Fishbourne Palace was indeed the residence of the Iron Age king Togidubnus, the presence of fallow deer at this site would suggest that the native elite were beginning to adopt Roman ideals, the ownership of game parks perhaps signalling membership of the Roman Empire

Whilst the specimens from Fishbourne Palace and Monkton provide conclusive evidence that fallow deer were maintained and bred in Britain during the Roman period, there is a big step from the localized breeding of exotic tame animals within elite parks and on off-shore islands to their establishment as a naturalized fauna. Currently, the evidence suggests that fallow deer died out with the withdrawal of the Roman Empire, being reintroduced in the late eleventh century (Sykes and Carden 2011).

As is shown in Figure 4.1, the medieval period saw a gradual re-emergence of trade in exotic animals, which appear with increasing regularity in the zooarchaeological record from the eleventh and twelfth centuries onwards (Pluskowski 2005; O'Regan *et al.* 2006; Brisbane *et al.* 2007). It is well known that exotic animals – especially big cats and monkeys – were frequently exchanged between leaders to cement political relationships. In all probability it was this symbolic significance of fallow deer, rather than their potential as quarry, that prompted their reintroduction to Britain. But it did not take long for the meat-hungry aristocracy to recognize and harness the venison-producing qualities of fallow deer. The aristocracy's desire to participate in the new culture of hunting and game-eating saw large numbers of fallow deer being imported and farmed to stock the burgeoning number of parks, which will be considered in more detail in the next chapter.

Summary

Animal introductions are frequently equated with the introduction of new dietary ingredients; however, none of the case studies presented above suggest that meat was the motivation behind the importation of these exotic species to Britain. Instead the evidence points towards a more ideological, or perhaps cosmological, inspiration for their introduction. It seems that most, if not all, of the species considered in this chapter were initially linked to deities: the horse with a human–equine divinity common, albeit in different guises, across Indo-Europe; the chicken with Mercury and Mithras; the hare with a less well-known female deity that may (or may not) have evolved into Eostre; fallow deer with Diana/Artemis and, although evidence for earlier introduction of domestic livestock is more obscure, Thomas (2007) and Cauvin (2000) argued convincingly that the spread of cattle and other early domesticates was also linked to new religious beliefs.

The connection of these new animals to the spread of new deities is important because different religions tend to be characterized by distinctive beliefs, social behaviour and ritual practices. Within sociology and cultural psychology it is widely accepted that religions and their associated practices serve as important mechanisms for structuring and controlling society (Graham and Haidt 2010; McCullough and Willoughby 2009; Saroglou 2011). Whilst today the arrival of a new religion may arouse suspicion and resistance (as we see with the Church of Scientology), in the past it is likely that new beliefs, especially if accompanied by new animals, would have taken primacy over existing traditions because they were introduced from beyond known realm and therefore, as Helms (1993) has argued, would have been viewed as more powerful cosmologically. In this way the arrival and establishment of domestic livestock, horses, chickens, hares, fallow deer and their associated deities are likely to have brought real changes for human ideology, society and social behaviour, impacting upon the ways in which identities and relationships were negotiated.

In some cases, the arrival of exotic species brought new opportunities and foci for socialization and community creation, as was seen in the case of cattle in the Neolithic and cockfighting in the Iron Age/Roman period. In other cases the new animals could become models for human behaviour and for gender definition, for instance cockerels and hens representing the ideals of masculinity and femininity respectively. It stands to reason that association with, or ownership of, exotic animals (e.g. the emparkment of fallow deer) would have conferred status and even bestowed power or authority, exemplified by the role of the horse within kingship rituals.

Importantly, the diffusion of new animals, together with their associated rites and behaviours, will have bound together communities from different regions, with practices being potentially understandable to groups across a wide geographical range. This is not to suggest a common identity but rather to highlight that animals are an excellent proxy for human and ideological diaspora. This is exemplified well by the mythology and rituals linked to horses, which are startlingly similar from Vedic India to early medieval Ireland: very different cultures separated in time and space but united by the horse. Whilst we can be certain that no direct connection existed between ancient India and Ireland, the same is not true for the Empires of the Romans and, later, the Normans. In both these cases exotica were used deliberately to link disparate territories, the animal connections becoming symbolic of a single powerful entity.

It is evident from the case studies outlined above that animal introductions characterize periods that witnessed considerable human mobility, be it in terms of physical diaspora, increased trade or ideological diffusion. Episodes of animal introduction mirror the rise and fall of hunting cultures and, as was outlined in Chapter 3, it seems likely that these fluctuations reflect wider shifts in worldviews and, in particular, attitudes to the natural world. Beyond ideological change, the introduction of domestic livestock, horses, hares and fallow deer must also have had a considerable impact on the way in which the natural world was physically negotiated and experienced, issues that will be considered in the next chapter on animals and landscape.

Chapter 5

IDEAS OF LANDSCAPE

Animals are, and always have been, central to the creation, use and perception of landscape. Human–animal interactions are the architects of the physical landscape, responsible for the form and location of settlements, roads, enclosures, woodland and fields. In other cases, animals may play a more psychological role in the construction of landscapes, their visual, audio and physical qualities providing media through which humans might experience and understand the world around them. Indeed, the meaning of a space is often defined, or at least evoked, by the human–animal interactions performed within it: maintaining domestic cattle within a field, chasing red deer across hunting grounds, or simply hearing the shrieks of seagulls at the coast. Whether as living organisms or as 'products' (meat, skin/fur, bones, artefacts), the behaviour and properties of animals – how they look, sound, smell, feel – are important ingredients for human cultural experience (Bartosiewicz 2003; Overton and Hamilakis 2013; Poole and Lacey forthcoming; Sykes 2010a). In sum, our everyday interactions with animals, whose lives are so intertwined with our own, inform the way we think about and behave in our environments; they give shape and meaning to our worlds.

The importance of animals in the creation of landscape is well recognized by cultural geographers, and there is a voluminous literature on human–animal geographies (e.g. Philo and Wilbert 2000; Wolch and Emel 1998). Despite this, archaeological studies of landscape have all too often removed animals from the equation, seeing humans as the only significant agents in landscape construction. Looking through the wider literature on landscape archaeology, animals are startlingly absent aside from their occasional incorporation into environmental and economic reconstructions (Ashmore and Knapp 1999; David and Thomas 2008; Johnson 2007; Rippon 2012). Recent years have seen an upsurge in more theoretically informed studies that emphasize the complexity of the cultural landscape and highlight the importance of human experience for the creation of place (e.g. David and Thomas 2008; Rogers 2007; Willis 2007). For instance, Moore (2007: 80) has argued that we can only begin to reach Iron Age perceptions of place if we examine landscape from its widest perspective and, in particular, give consideration to material culture – the theory being that objects give a space meaning. I concur with Moore's view and would argue that the same case can be made, but even more strongly, for animals – living things that have the capacity to form two-way relationships with humans in a way that artefacts do not.

Not all zooarchaeologists appear to be of the same opinion: Mainland (2008: 546) infers that faunal remains offer little beyond palaeoenvironmental characterization, stating that 'with the advent of animal domestication and the subsequent dominance of domesticates in many archaeological assemblages, the potential of faunal data to provide insights into past landscapes is much reduced'. I do not subscribe to this perspective, quite the opposite, and believe that recent zooarchaeological studies (e.g. McCorriston *et al.* 2012; Pluskowski 2007; Pluskowski *et al.* 2011; Sykes 2007c, 2010a) are beginning to move the discipline of landscape archaeology forward (see in particular Creighton 2009; Fleming 2010).

Despite these advances it is important to continue stressing and demonstrating the role of animals in landscape experience and perception. To this end, this chapter seeks to put animals back into the landscape. It takes as it central tenet the idea that animals (whether alive, as carcasses or as products/foodstuffs), landscape, social structure and ideology are inter-connected to the point that a change in one will be reproduced in the others. In this way, human–animal–landscape interactions are culturally specific but this means that, if we can come closer to understand the relationship, there is potential to gain new insights into the worldview of the culture under study.

What Is Landscape and Why Study It?

Landscape is perhaps one of the most nuanced and contested terms in archaeology and therefore it is only right to begin with some definitions. The word 'landscape' is actually a fairly modern construct: it entered the English language *c.*1600 AD when it was imported from Holland along with the artistic tradition of landscape painting. The term comes from the middle Dutch 'landscap', meaning *land* (land) and *scap* (creation or condition) but cognates are found in Old English (*landscipe*), Old High German (*lantscaf*) and Old Norse (*landskapr*), the last three referring to human-made bounded tracts of land.

It is perhaps this association – that landscape is seen as something physically shaped by humans – that has been most influential within archaeology. As a human construct, landscape can be viewed as the biggest artefact available, with the potential to be read as a document onto which successive generations have inscribed their histories and identities (Ashmore and Knapp 1999: 18; Aston and Rowley 1974; Roberts 1987). The belief that the landscape can be 'read' has been responsible for a predominance of mapping excercises, with archaeologists striving to trace the features and settlement patterns that survive in the landscape. However, as David and Thomas (2008) have highlighted, this focus on mapping and environmental/economic reconstruction has encouraged the landscape to be viewed as a static backdrop onto which human behaviour was performed, rather than being the dynamic nexus of human experience, daily life and beliefs.

In this chapter, I define landscape not simply as physical geography, environmental conditions or human constructions, although these are all important elements, but also as how people perceive and understand their surroundings as a result of their engagement with them. This definition sees landscape, daily practice and worldview as inseparable, each serving to constitute the other. Pitts (2007: 698) has suggested that landscape evidence is 'able to provide only a low-resolution approach to change in everyday practice' and perhaps this is the case where attention focuses on mapping landscape features. However, the approach advocated here is that animal remains can be examined as direct reflections of human–animal–landscape interactions and thus provide clear insights into daily practice, especially when integrated with other sources of evidence. If it is accepted that interactions with animals are formative to human experience and are key to the creation and perception of landscape, then animal remains – and the human–animal–landscape interactions that they represent – can be used to examine the ways in which people engaged with, traversed, and so comprehended, their surroundings.

Animals, People and Landscape

The fact that I have to make a case for the relationship between humans, animals and landscape is a reflection of our time and worldview. In the past this tripartite inter-connectivity was more widely acknowledged and is demonstrated by a variety of evidence. Starting with most recent and going back through time we can look first to maps and landscape illustrations – unlike modern examples, which show pristine landscapes eerily devoid of the biota that inhabit them, earlier cartography provide not only depictions of physical geography but also the animals (and plants) found therein (Figure 5.1). According to Kline (2001: 98) animals were sprinkled across medieval maps like 'confetti' and the same can be said of place-names, many of which contain animal elements (Aybes and Yalden 1995; Gelling 1987; Yalden 1999). From this it is clear that, since at least the medieval period, animals were an important component of the psychological landscape. This is perhaps unsurprising given that before the Enlightenment, when humoural principles dominated thought (Chapter 1, page 18), the place of animals mattered. It was believed that humans could acquire the temperament of animals via any of their senses – touch, smell, aurally but also sight. For instance, an early thirteenth-century document, attributed to Robert Grosseteste, explains that it was considered sinful to see animals copulate and that the bodily pollution incurred through the vision of such spectacles would need to be confessed (Goering and Mantello 1986). The sound of animals could be equally corrupting, the song of the nightingale famously being said to induce lust in those who heard it (Leach 2006). With animals having the potential to engender such mortal sin, it is understandable that people may have been keen to know their whereabouts.

Medieval maps not only illustrate animals but, in some cases, they suggest that animals were directly responsible for the form and layout of the cultural landscape.

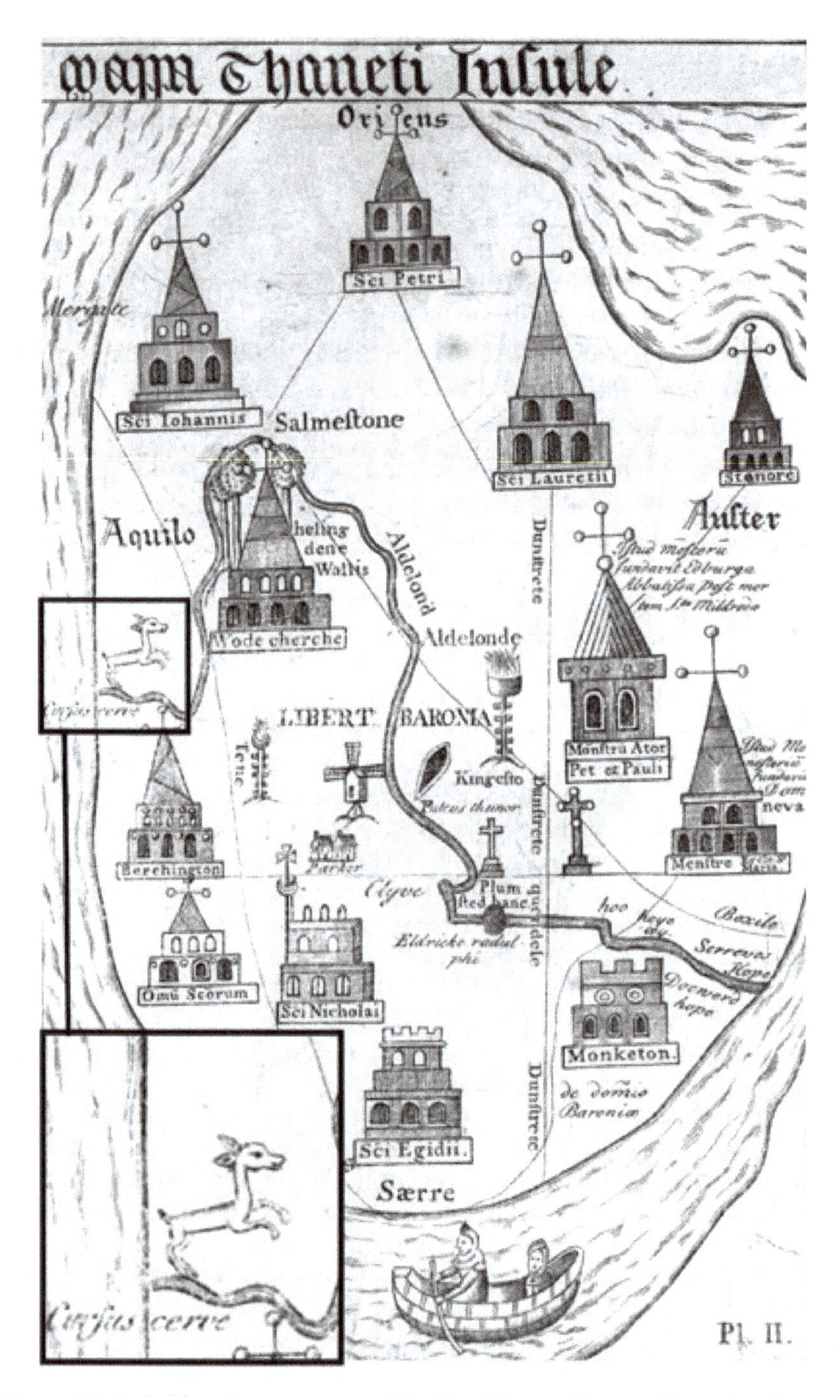

Figure 5.1 Early fifteenth-century map of the Isle of Thanet. The erratic boundary line running from the left to right of the island is said to have been laid out by the pet deer of the seventh-century princess Domneva.

For instance, Figure 5.1 shows a fifteenth-century map of the Isle of Thanet (Kent) that depicts the boundary of the island's minster estate which, according to legend, was laid out following the wanderings of the pet deer owned by the seventh-century Kentish princess Domneva (Hiatt 2000: 868). A similar story is found in Roman origin myths, where the boundaries of the city state of Rome are said to have been delineated by cattle and that the position of the city's walls and gates were marked out by a bull (Schwabe 1994: 46). Indeed, the very name Italy was said to have been bestowed because of the number and beauty of its cattle, the name 'Italy' coming from the ancient Greek for cattle, '*itali*' (Schwabe 1994: 46). In these few lines the connection between people, animals and land is made explicit; that people and animals work together to create cultural landscape and cultural identity.

In the case of both Rome's boundaries and Domneva's estate, animals played a central role in the demarcation of territorial rights and proprietorship. In many societies, however, the concept of land ownership does not exist in quite the same way; thus fixed, continuous property boundaries should not always be expected (Landais 2001: 474). Nevertheless, Perrin (2011: 627) has argued that humans have an innate predisposition for environmental marking and suggests that they have always done so, in much the same way that other animals (e.g. bears, wild boar, large cats and deer) score marks on the landscape, in order to signify boundaries, territorial distribution or rallying points.

Perrin (2011: 625) suggests that, prior to the invention of land ownership, animal tracks would have been a vital component for human perception of landscape. This is particularly the case amongst hunter-gatherer groups whose lifestyle and patterns of mobility are intimately linked with the movement and migration of the animals that they follow (Chapter 2; Ingold 2000: 67; Willerslev 2007: 42). Under such circumstances landscapes are constituted and perceived through specific and repeated human–animal interactions that occur at particular times and places. Very often these relationships are woven into human culture, being preserved in landscape features, place-names, art, songs, dance and stories. A good example of this is provided by Sommerseth's (2011) work on the reindeer hunters/herders of Norway. In this region extensive chains of pit-fall systems for hunting reindeer were laid out around 2600 to 2000 BC, and rock art depictions of corrals suggest that landscape features associated with reindeer hunting/management date back to at least 5200 to 4200 BC. Sommerseth (2011: 116) argues that these structures indicate that the people responsible for their construction had a detailed knowlege of reindeer ecology, migration routes and terrain that must have been passed from generation to generation. Indeed, evidence for this is perhaps provided by later Sami place-names, particularly those found along historically documented reindeer migration paths. These place-names reference land formations, ritual practices, past events and mythology associated with reindeer hunting. In this way the landscape can be seen as a 'living narrative' that connected people, reindeer and place, and served as 'topographic tools' to help those involved in long distance movement visualize and navigate their path (Sommerseth 2011: 116).

Returning to the idea of territoriality, it is possible that the reindeer themselves served as markers, representing dynamic landscape boundaries that shifted with the herd's movement. Whilst this concept may seem very alien, it would go some way to explaining the practice of animal marking – e.g. painting, shaving or fire-branding the skin, or clipping of the ears, horns or antlers – which has been observed not only amongst the Sami (Sommerseth 2011: 117) but across cultures worldwide, some dating back to the Neolithic (see Figure 5.2a and Landais 2001). Perrin (2011: 625) has proposed that the human marking of animals is a form of territorial behaviour that probably dates back to the Palaeolithic, and he provides examples from cave art to support his case. Whether or not the practice has such great time-depth is debatable but wherever evidence exists – be it in the form of ethnography (Dransart 2002; Landais 2001; Stammler 2012), historical documents and place-names (Sommerseth 2011), rock art and archaeological remains (Sundkvist 2004) or a combination of all of the above (Chamberlain 2005; Dransart 2002; Russell 2013) – the data suggest that the marking of animals is, and has always been, widespread. Sadly, traditional zooarchaeology is a poor medium for the detection of such behaviour due to issues of soft-tissue preservation but, on rare occasions, some glimpses of the practice can be seen, such as in the case of the ear-notched horse mummies from Siberia (Figure 5.2b).

Despite the apparent ubiquity of animal marking, the meaning and function of the practice varies through time and between groups (Landais 2001: 477; Russell 2013). Today we tend to perceive animal marking in terms of signifying ownership but this is a reflection of a worldview that sees animals as commodities. In many other cultures, such as the Tuareg of Africa, the Bara of Madagasca, the Mongols of Siberia and Mongolia and camelid herding groups of the Andes, marking is employed to show that an animal is *part* of a clan rather than being its *property* (Dransart 2002: 96–100; Landais 2001: 473; Russell 2013: 4–7; Stammler 2012). For these peoples, animal marking can be understood in much the same way as human scarification practices, as a rite of passage; indeed, there are some modern groups where both humans and animals carry the same clan marks (Landais 2001: 473; Russell 2012: 8). Other reasons for marking include identification (of both animals and peoples) as well as magico-religious beliefs, the idea being that marking will protect animals from theft and disease, issues that will be considered further in Chapter 6. These multifarious reasons need not be mutually exclusive: for instance, when Chamberlain (2005) interviewed members of the Samburu of northern Kenya, respondents gave multiple and occasionally contradictory replies about the rationale for animal marking. However, the one theme that occurs repeatedly in the anthropological literature pertaining to animal marking is that it is linked to perceptions of landscape.

Landais (2001: 474) has suggested that it may be inappropriate to refer to the practice as 'livestock' marking because, in many instances, the emblems are also drawn on a variety of personal possessions (knives, weapons, chests, stools) as well as gravestones, trees, rocks, dolmens and menhirs. Amongst many pastoral groups this practice is explained as a mechanism by which a clan asserts its rights over resources, such as grazing land or watering points. However, this is not to suggest

Figure 5.2 (a) Neolithic (*c.* 3000 BC) fresco from Ouan Derbaouenm, showing bovine carrying mark on cheek, similar to those used in the region today (after Landais 2001); (b) Ear-marks of horse recovered from the Iron Age (*c.* sixth to third century BC) Scythian graves of Pazyryk (after Sundkvist 2004).

that the clan 'owns' the resources as many groups, descended from a single ancestor, will share the same rights (Landais 2001). Instead, Landais argues that the application of the ancestral clan mark to people, animals and geographical features serves to unite communities (living and dead) with territories, enabling all to be organized and managed in social, spiritual and economic terms.

The practice of livestock marking demonstrates the central role that *living* animals can play in the creation and perception of landscape; however, their death can also serve to constitute landscape. In their studies of African pastoralist groups Chamberlain (2005) and Russell (2013) both highlighted a direct correlation between the location of rock art depictions of clan marks (and other emblems) and feasting sites. Through interviewing the Samburu, Chamberlain (2005) was able to ascertain that these locations were visited specifically to eat meat, which seldom occurs at the homestead, and that the clan marks were made to show that the participants had eaten their own animals. Interestingly, it is tradition amongst the Samburu to paint the motif using the fat from the slaughtered individual. In this way, the animal becomes part of the cultural landscape, the painted mark a physical mnemonic for the slaughter and consumption experienced at the site, whilst also signifying to others that members of the clan are in the area.

This practice of feasting at particular points in the landscape is reminiscent of the situation proposed for the cattle feasts of Neolithic Britain (Chapter 4, page 80) and similar conclusions have been drawn by McCorriston *et al.* (2012) relating to their findings from Neolithic Arabia. McCorriston *et al.*'s (2012) study focuses on a unique set of forty-two domestic cattle skulls, recovered from Shi'b Kheshiya dated to the fifth millennium BC, that were pushed nose down into soft mud to form a monumental skull circle (Figure 5.3). The manner in which the circle was constructed, with the animals' horns interlocking, indicates that it was the result of a single event. Furthermore, the association of the circle with a stone platform and numerous hearths suggests that the animals were slaughtered at the site, with the resulting surfeit of meat being cooked to feed a large number of individuals. The authors argue that the site can be interpreted as a form of social territoriality, or 'social boundary defence', whereby access to pasture and water resources was negotiated through cattle sacrifice and feasting. Attendance at the feast, presumably requiring the donation of a sacrificial animal, would demonstrate that a group was part of the wider community/territory and as such had rights to the resources available.

Similar explanations may be proffered to account for other landscape monuments associated with feasting, such as the Siberian 'deer stones' where horses were sacrificed and eaten (Fitzhugh 2009), the shell-mounds of Brazil which are composed of both feasting debris and funerary deposits (Okumura and Eggers 2010), and the Bronze Age middens of southern England (Yates 2007).

Feasting as a form of social boundary defence need not, of course, result in the creation of a monumental structure within the landscape. The animal itself may be the 'monument', the division and sharing of its meat representing the landscape. For example, amongst the Nootkan on the northwest coast of North America the whale is of exceptional importance and on occasions, when one has been hunted,

Figure 5.3 The Shi'b Kheshiya cattle ring.
Source: Joy McCorriston.

its body becomes a map of social and territorial relations. Busatta (2007: 5) quotes the words of a Nootkan chief who stated that 'when the whale was cut it represented every inch of our chief's territory, every cut had to be precise. You could not cut into another chief's portion because that meant part of his territory was being cut off.' The whale meat was then distributed according to the rights and rank of the recipient, a practice that Busatta (2007: 4) highlights is similar to the traditions of meat redistribution witnessed in medieval Europe.

For early medieval England, I have argued for this connection between meat, land and social hierarchy through a study of hunting and venison redistribution (Sykes 2010b), which will be considered in more detail in Chapter 8. For the remainder of this chapter, I will stick to the theme of hunting landscapes and use the example of the park – an icon of the English landscape – as a short case study to demonstrate how animals can provide new perspective on old questions.

The English Park: Origins, Meaning and Function

English parks provide a prime example of where mainstream landscape research has neglected animal studies; landscape historians and archaeologists traditionally placing more emphasis on the reconstruction of park boundaries than on the animals and activities that occurred within the pale (Cantor and Wilson 1961). More recently, the economic and social functions of parks have started to be recognized, with enclosures described variously as game larders (Birrell 1992), masculine hunting spaces (Gilchrist 1999: 145) and socially divisive symbols of power (Herring 2003; Liddiard 2000). Without giving detailed consideration to the meaning of, and human interaction with, the animals that these enclosures contained, however, it seems difficult to elucidate their function and social significance. In the absence of this information it is not possible to know when the concept of emparkment emerged and if or how the meaning of wild animal enclosures changed through time.

Park Origins

Traditionally the origins of the medieval park have been attributed to the Norman introduction of fallow deer and, as was seen in Chapter 4 (page 96) their establishment and population increase is clearly associated with the burgeoning of emparkment. However, given the recent discovery of fallow deer remains in Roman Britain, in particular at Fishbourne Roman Palace in Sussex, it is possible to push the origins of the deer park further back in time. The significance of the deer park, or *vivarium*, at Fishbourne is considered in Chapter 4 (page 96), where it is suggested that it symbolized sanctity, social standing and Imperial membership and allegiance. In terms of landscape meaning, this is probably quite similar to that of the earliest Norman parks, which probably served more as menageries for exotica, perhaps validating their oft-cited classical origins (e.g. Rackham 2000: 122). Most importantly, the evidence from Fishbourne demonstrates that even where physical evidence for parks cannot be discerned, their presence can be demonstrated through zooarchaeology. By the same principle it may be argued that zooarchaeology can also highlight when parks were not present in the landscape – for instance, there is no evidence for fallow deer in Anglo-Saxon England, suggesting that parks too became locally extinct following the withdrawal of the Roman Empire.

Liddiard (2003) has argued that the Anglo-Saxon landscape did contain deer parks. Certainly Hooke (1989) has shown that some place-names including the elements *haie* (hedge) or *haga* (enclosure) refer specifically to deer, such as the *derhage* (deer enclosure) at Ongar in Essex (Rackham 2000: 125). The fact that Ongar's *derhage* later became part of Ongar Great Park has led several researchers to suggest that *haie, haga* and *deor-fald* were little different in role and physical form to post-Conquest parks (Hooke 1989; Liddiard 2003). There are many reasons to suspect that this assessment is correct but, again, it is worth considering the animals these structures actually contained if we are to understand their meaning.

As was seen in Chapter 3, the zooarchaeological record provides little evidence for hunting in Anglo-Saxon England although during the later part of the period (late ninth to late eleventh century) elite sites begin to demonstrate a higher frequency of wild animal remains. In nearly all cases, roe deer (*Capreolus capreolus*) are the dominant species, particularly within elite assemblages (Sykes 2007b). The abundance of roe deer at the high-status site of Faccombe Netherton in Hampshire is interesting given that a 'white haga' was recorded here in AD 961 (Hooke 1989: 128). Documentary evidence also suggests an association between *haiae* and roe deer: the Domesday survey for Cheshire and Shropshire mentions *haiae capreolis*, hays specifically for roe deer (Thorn and Thorn 1986: 6.14, fol. 260b).

If, as it seems, roe deer were the main quarry in these wooded landscapes, the interpretation of *haiae* and *hagan* as parks is problematic. This is because roe deer are notoriously unsuitable as park animals, the males in particular becoming dangerously territorial when confined. Liddiard (2003: 20) has pointed out that early English parks tended to be fairly large, which may have alleviated the problem to some extent. However, I would suggest that the *haiae* and *haga* associated with woodland were most probably intermittent boundary structures rather than continuous enclosures – thus enabling animals to enter and leave the woodland. This would explain why medieval documents, such as the Domesday Book, often record *haiae* in the plural, whereas parks are referred to in the singular (Liddiard 2003: 12).

The Anglo-Saxon hunting of roe deer within woodland, an environment thought to be the residence of supernatural spirits as well as wolves and other dangerous animals (see Chapter 3, page 69), must have placed the hunter mentally alongside the warriors, heroes and saints of Old English literature (Neville 1999). However, would medieval parks, stocked as they were with exotic fallow deer, have been perceived and experienced in the same way? It seems unlikely and I would argue that even where there was continuity in physical space, such as with the *derhage* and later park at Ongar, the pre- to post-Conquest difference in human–animal interaction rules out any possibility of continuity in landscape meaning. For this reason, I would place the origins of the medieval deer park squarely with the introduction of the fallow deer in the late eleventh century (Chapter 4, page 96)

Park Function

Fallow deer were not the only animal species contained within medieval parks. By the later medieval period they were deliberately stocked with a wide range of animals: rabbits (*Oryctolagus cunniculus*), pheasants (*Phasianus colchius*), peafowl (*Pavo cristatus*), partridges (*Perdix perdix*), swans (*Cygnus* sp.), herons (*Ardea cinerea*) and freshwater fish.

Whilst the function of medieval parks was clearly multifaceted (Liddiard 2005) they are, today, most frequently perceived as private spaces for hunting. Certainly their stock and design would have offered opportunities for a bountiful hunting experience that could not be achieved in the wider landscape; but they also imposed limitations, requiring the park to be engaged with and traversed in very particular ways. With this in mind, it is surprising that few scholars have examined how parks could have operated as hunting spaces (Cummins 1988). Within a park stocked with fallow deer, hunting methods must have been limited to the drive, or 'bow and stable', whereby deer were chased into nets or towards archers who were positioned on a platform, or 'stand'. This is indicated not only by historical records but also by the zooarchaeological evidence: arrow fragments have been found embedded in fallow deer remains (Sadler 1990: 487).

Based on Cartmill's (1993: 30) criteria for hunting (see Chapter 3, page 52) it is debatable whether park-based bow and stable can be considered as 'hunting' at all. It is certainly possible to take issue with concepts of wildness, freedom, and direct violence in a situation where, although known as 'wild animals' (*ferae*), fallow deer were in effect farmed, restricted and killed from a respectable distance (Birrell 1992: 115). Park 'hunting' can have done little to stir warrior-like emotions and the general dearth of historical evidence for seigneurial hunting within these landscapes suggests that it occurred infrequently as a lordly activity (Birrell 1992: 122). There is equally little evidence that lords engaged in rabbit hunting, which according to (Cummins 1988: 236–7) was an unsuitable activity for respectable lords.

Whilst it may not have been fitting for a nobleman to participate in rabbiting, medieval illuminations suggest that this was an appropriate pursuit for a lady: see, for instance, Queen Mary's Psalter (BL MS Royal 2B VIII, f.155) which shows ladies taking rabbits with ferrets. Its suitability may have derived from the fact that it required minimal physical exertion and that, even in the medieval mind, rabbiting was not classed as true hunting. Similar reasons have been put forward to explain the association between falconry and medieval women: ladies were able to maintain their femininity as they were not responsible for the kill, which was mediated by the hawk or falcon (Almond 2003: 160). Considering that falconry was a well-known part of a young woman's education, little thought has been given to establishing where in the medieval landscape such activities might have taken place. Gilchrist (1999) has argued cogently that medieval courtly society placed great emphasis on the seclusion and enclosure of young aristocratic women, seeing this as a way of ensuring the female chastity and purity so essential to patrimony. With their mobility and visibility restricted in public arenas – parish churches, banquets and even travel between households – it seems highly unlikely that young women would have been encouraged to practise falconry openly in the wider landscape. More probable is that it was within the confines of the park that ladies, such as the unaccompanied female shown in the Taymouth book of hours (BL MS *Yates Thompson* 13 f.73), engaged in falconry without fear of observation.

It makes sense that parks were arenas for falconry, being stocked as they were with the very species – the heron, partridge and pheasant – that were amongst the sport's most prized quarry. It is notable that the zooarchaeological representation of these species increases dramatically on elite settlements dating to the mid-twelfth century (Sykes 2004), a shift coincident with an upsurge in iconography, particularly ladies' seals, depicting females with hawks (Oggins 2004: 118). Perhaps it is not unfeasible that these prey species were managed in parks specifically with women's leisure in mind. If it is accepted that parks were the setting of women's falconry, the natural step in the argument is to suggest that they may also have been the location for ladies' deer hunting.

The level of female involvement in medieval hunting has been discussed by several authors (Almond 2003; Cummins 1988). Controversy has stemmed from the numerous manuscripts that depict women hunting or unmaking animals: do these illuminations reflect real practice or are they simply part of the 'topsy-turvey world' of artistic tradition? Some of the images are clearly comical but others appear more realistic, depicting women shooting arrows at rabbits and deer: significantly, where women are shown in association with deer it is often with the fallow deer, the principle park inmate (Richardson 2012). Whilst it is unwise to read medieval images at such a superficial level, clear evidence that women hunted within parks can be found in the historical record. For instance, in John Coke's *Debate between the Heralds*, the English herald states: 'we have also small parkes made onely for the pleasure of ladyes and gentylwomen, to shote with the longe bowe, and kyll the sayd beastes' (Cummins 1988: 7). To conclude that parks played a role as hunting spaces for women may seem to be at odds with my earlier argument that deer-killing within parks is difficult to classify as hunting. However, it is this very fact – that standing at a station and shooting at captive deer is little more than target practice – which made the activity a suitable one for women: it was the masculine arrow, rather than the holder of the feminine bow, which caused the death, and the kill was sanitized by physical distance.

When the evidence is viewed together, the impression given is that, for the aristocracy at least, parks functioned as extensions of the garden. Beyond landscapes of animal production they were social grounds where young ladies could engage unobserved in sporting activities and where, under the pretext of 'hunting and hawking', noble men and women could interact: as with the garden, the park was a contested space (Gilchrist 1999). Later medieval parks clearly had multiple functions and must have been viewed in equally diverse ways. It is now worth considering whether animal studies can shed light on the varied meaning of park landscapes and, in particular, the ways in which parks were perceived by different social groups.

Perception and Meaning

Parks and the features they contained – warrens, dovecotes and ponds – are frequently described as symbols of social division: their boundaries a physical reminder of the cultural partition between aristocracy and peasantry. To their owners, parks might have been viewed as landscapes of pleasure and plenty, but for others their connotations must have been very different. Herring (2003: 46) sums up popular ideas about peasants' perceptions of parks with his suggestion that, to them, parks represented their 'powerlessness; lowliness of rank; and separateness or exclusion'. Certainly, the strict protocol of park management and game distribution was established to preserve ideals of exclusivity; park boundaries were maintained at great expense and the distribution of game meats was tightly controlled. But it was the creation of this very institution that provided excluded individuals with both an opportunity to subvert authority and a target for expressions of defiance, hence the spate of park breaking during the Peasants'

Revolt of 1381 (Liddiard 2005: 118). Our image of the depressed and powerless lower classes is perhaps underestimating the spirit of the average medieval peasant. That many of the lower classes viewed parks not so much as an oppression but rather as a challenge is clear from the level of poaching recorded in the documentary sources (Birrell 1982, 1996; Manning 1993). We should not see poaching as a peasant's last resort, something they did to avert starvation – there is very little evidence that peasants poached because they were hungry (Manning 1993: 20). Rather, parks were breached as a social statement. In many cases peasants and nobles poached together, particularly in cases of inter-household feuds, where aristocrats launched raids on the parks of their rivals (Manning 1993). Under these circumstances, far from being socially divisive, parks were places where the gap between the classes could actually be narrowed.

Liddiard (2005: 110) has highlighted the religious symbolism of park features, and it is not difficult to find Christian imagery associated with park animals. Stocker and Stocker (1996) have argued that rabbits were maintained not simply for their socio-economic value but for metaphorical reasons: their fecundity, subterranean habitat and surface emergence were viewed as an allegory for human life, death and salvation, respectively. The same may be true of deer, which were repeatedly invoked as symbols of Christ, eternal life and resurrection (Cummins 1988, 68). Herons were similarly well regarded, being linked to Christ and piety (Rowland 1978: 80). It seems possible, therefore, that these park animals were chosen specifically for their Christian connotations. A further Biblical dimension was then added through the act of caring for them: stewardship of animals was seen as a sign of sanctity, fulfilling the duty of care (*dominion*) towards creatures set out in Genesis 1, 26 (Preece and Fraser 2000). In this way the meanings of parks went beyond mere expressions of socio-economic status; they represented spirituality, anchoring their owner at the highest position in the Chain of Being. Perception of parks as sacred landscapes may explain why so many were established and owned by members of the Church. The number of parks in archiepiscopal, episcopal or monastic hands has struck some as odd, especially given that hunting was forbidden by canon law (Coulton 1925: 508–12). However, the dichotomy seems less severe if it is accepted that parks were not true hunting spaces – animals may have been killed but only through passive methods.

At the other end of the moral scale, parks might contain animals that were the very essence of profanity. According to Rowland (1978), both the pheasant and the partridge had strong sexual connotations. The partridge was seen as particularly wanton and impure, a symbol of 'incontinent lust' (Rowland 1978: 124), an association which appears to have removed this species from the parks and tables of religious houses (Sykes 2004: 96; see Chapter 8). Similarly, whilst the activities that took place within parks were seen as pure on the one hand, they could be eroticized on the other, unmaking scenes being classic devices for portrayals of courtly love (Almond 2003: 154). This interplay between the sacred and the sexual is similar to the conflicting imagery that has been identified for the castle garden: an enclosed feminine space that, like women, was seen as a symbol of both sanctity and lust (Gilchrist 1999). That we see the same type of ambiguous motifs represented within parks strengthens the argument that these enclosed spaces were also strongly feminized.

Summary

Today we give little attention to the place and significance of animals in the modern landscape but the same was not true in the past – there is clear evidence that animals were central to the physical and psychological lay of the land. This chapter has sought to demonstrate how, as living creatures, the presence, behaviour and markings of animals can influence the way in which humans perceive and understand the world around them. In death, animals need not lose their significance: their carcasses may symbolize the landscape in total or their body parts may become incorporated into physical space as markers of culturally important places.

The essential message that I have tried to convey is that human–animal–landscape relationships articulate the ideology of those involved in their construction. These ideologies can be culturally overarching whilst simultaneously specific to individuals. For instance, I have argued in this chapter that later medieval parks were multi-dimensional spaces more than mere status symbols or providers of game meats; they were an important structuring force that operated concurrently to both amalgamate and segregate the different classes and genders of medieval England.

While this chapter has taken a rather wide-ranging approach, I hope it has demonstrated that animal studies, in particular zooarchaeology, can offer new insights into old questions – such as those relating to the well-trodden ground of English parks – and have the potential to play a central role in landscape research and wider cultural interpretation. However, if landscape is a cultural nexus it must be expected that it is implicated in everything that we study and cannot be considered in isolation. To some extent, landscape has already been invoked in preceding chapters (notably Chapters 3 and 4) and it will continue to be referred to throughout the rest of the book as it cannot be separated from issues of food (Chapter 8) or cultural identity (Chapter 9). Similarly, as has already been seen in the case of animal marking, the Shi'b Kheshiya cattle ring, and even medieval parks, it is difficult to separate landscape from wider belief systems. And it is those belief systems that we label as 'ritual', that will be considered in the next chapter.

Chapter 6

THE CHAPTER ABOUT RITUAL

We all love a bit of 'ritual' – stories of gory sacrifice and weird human behaviour that re-awaken our attention when we have drifted off during discussion about farming regimes. Who cares about dull domestic rubbish when we can find out about exciting 'special deposits'? Down with the boring economy! Up with social analyses of ritual! Right? Well . . .

I hope that by now, if nothing else, this book has demonstrated that life can seldom be so neatly compartmentalized into ritual or functional, social or economic, and religious or secular. Western culture with its love of bureaucracy, classification and human–animal separation is poorly placed to understand this complexity, which is probably why most zooarchaeologists tend to deal with animal-associated ritual as a distinct behavioural category, identified primarily when animal remains are recovered from contexts that appear to represent activities outside the day-to-day (Groot 2008). There is also a tendency for zooarchaeologists to concentrate only on rituals associated with animal death, with particular emphasis placed on the identification of sacrificial killings (e.g. Pluskowski 2012; Russell 2012: 88–126). Whilst this focus on dead animals is entirely understandable for a discipline that deals primarily with the remains of dead animals, such an approach overlooks all those lifetimes of interaction that would have provided ample opportunities for a wide variety of rituals to have been carried out.

There are many anthropological and historical examples of animal-based rituals, and even 'sacrifice', where the prerequisite is that the animal is *alive*. For instance, in West Java there is a tradition of offering live chickens to the *siluman*, a kind of ambiguous spirit, who is said to inhabit the *Bumi Buana* (hidden world or place) of the local hills. According to Wessing (1988 and personal communication) these 'forest chickens' (*ayam hutan*) are released to roam the hills as the *siluman's* livestock. Today, live chickens are gifted by people in the hope that, in return, the *siluman* will tell them the winning lottery numbers: no better example of the intertwining of 'economic' and 'ritual'.

Anthropological and historical evidence suggest that in most societies outside modern Western culture, animal-associated rituals are everywhere, worming through all aspects of human and animal life, not just their death (see Abbink 1993; 2003; Crate 2008; Landais 2001). In this chapter I will use the life-cycle as a device to consider different forms of animal-related ritual and how they might be detected (or not) in the archaeological record. Before this, however, we will consider what the term 'ritual' might actually mean.

Definitions of Ritual

One of the most often-cited archaeological definitions of ritual is that it is an "all-purpose explanation used where nothing else comes to mind" (Bahn 1989). This definition was, of course, both tongue-in-cheek and a product of its time, when ritual interpretations were being heavily critiqued (Morris 2008a). Whilst attitudes have moved on, the idea that archaeological studies of ritual are subjective interpretations of actions with intangible motivations has endured to a large extent (Howey and O'Shea 2006). This has seen ritual continue to be perceived as non-functional or non-economic irrational behaviour (Brück 1999). As a result, archaeological research into ritual has been limited, with the exception of studies undertaken within the context of religion (e.g. Insoll 2011). Indeed, there is a plethora of zooarchaeological publications that focus on rituals within funerary contexts (Bond and Worley 2006; Brück 2006a, 2006b; Gräslund 2004; Hamilakis 1998; Pluskowski 2012) and at religious centres (Hamilakis and Konsolaki 2004; King 2005; Pluskowski 2012; Popkin 2010). However, must ritual always be equated with religion? Is it not possible that rituals can occur within secular settings? For example, many traditions of hospitality – such as the etiquette surrounding the seating and serving of diners, or the portions of food and drink that guests receive – involve rituals that appear more social and/or political, than religious (see Chapter 8; Gautier 2009; Lokuruka 2006). If ritual can occur within any context the question remains, what purpose do they serve?

Groot (2008: 99–105) provides one of the best summaries of the anthropological and archaeological literature pertaining to ritual. Quite sensibly, she gives up on the task of actually defining ritual, explaining that there are as many definitions as there are researchers. Instead she usefully offers a number of broad traits commonly identified in manifestations of ritual behaviour. Notable amongst these is the element of empowerment, often through communication with the supernatural, whereby people try to strike reciprocal deals with deities or other supernatural forces in order to achieve a particular goal – be it fertility, success, safety, wealth, health or well-being.

Such communications may take place in special locations and there is often an element of public display; even when rituals are conducted in private there will be a supernatural audience (Groot 2008: 100). Repetition, formality and rules also tend to be key components, and it is these elements that render rituals such a useful mechanism for community cohesion: they help to strengthen, structure and maintain social identity and order (see Chapter 8). This makes clear that ritual is far from irrational; it fulfils an important role at the level of society. However, rituals are also psychologically valuable down to the level of the individual. In a world that is, essentially, chaotic, it is soothing to believe that we can get the supernatural 'on side', the idea serving to calm anxiety about the uncontrollable unknown (Groot 2008: 103).

As with most zooarchaeological discussions of ritual, Groot (2008) succumbs to the tradition of concentrating on the dead, rather than the living, animals. Overton and Hamilakis (2013: 113) have argued that this approach defeats the aim of 'social zooarchaeology' because rather than moving away from economic interpretations, it presents ritual as just another product for which animals can be exploited. In order to challenge the idea that animals are passive ingredients of cultural rituals, this chapter will now examine how animals have played central roles in human life and death, with animal-associated rituals being at the very heart of human culture, even having the capacity to influence human decision-making. Inevitably I will examine animal sacrifice and the significance of dead animals but, in a break with tradition, I will start by focusing on living animals and their role in life-cycle rituals.

Live Animals and Rituals of the Life-cycle

In many modern non-Western cultures the lifecycles of humans and animals are intimately linked, as they walk hoof-in-hand-in-claw through their respective existences. I have already tried to make this point with regard to the Suri of Africa where human–cattle partnership is central to the life-course of both (Chapter 1 and 2, see also Abbink 1993, 2003). Another of my favourite examples derives from Wessing's (2006) work in Southeast Asia where, as with the Suri, animals and humans are connected from the very start of their lives, which is where we shall begin.

Birth and Coming-of-Age Rites

Across Indonesia the human spirit is thought to take the form of a chicken; indeed, if a person is feared to have lost their spirit, it is called back to their body with the chant 'kurr! kurr!' traditionally used in chicken herding (Wessing 2006: 223). When a child is born, an important part of the post-partum ceremony is the giving of a young domestic fowl called the *hurip*, meaning 'alive' (Wessing 2006: 221). According to Wessing (2006) the welfare of the chicken and the baby are deemed to be connected and therefore the mother of the newborn human cares equally for both juveniles. It is true that many chickens meet a sticky end as a result of ritual practices in Indonesia (see below); however, for this particular ritual, the important point is that the *hurip* chicken is very much alive and continues to be so. Assuming that the human individual survives, the chicken does likewise and eventually returns unharmed to its flock (Wessing personal communication). Such rituals of human–animal connection would, therefore, be archaeologically invisible where the human lives successfully through the first few months after birth. However, discoveries of human–animal co-burial (such as the infant/piglet burial at Neolithic Tell 'Ain el-Kerkh, see Russell and Düring 2006) could hint that similar relationships and rites existed in the past, preserved archaeologically only where the human child dies in infancy.

Assuming that the human infants survive, we might expect that further rituals are conducted to ensure their well-being. And such rituals need not be restricted to humans. Amongst the Saka of Siberia, animal-directed rituals begin at birth with particular practices being undertaken to placate the *Inakhsyt–Inakhsyt* (God of Cows) and ensure that calves are healthy and obedient (Crate 2008: 119). To achieve this, the spirits are 'fed' by lighting a fire of dry manure and hay, onto which is put milk and cream as well as pancakes made of the colostrum-rich milk that the cow produces immediately after calving. The act of burning allows the *Inakhsyt–Inakhsyt* to partake in them via the smoke that is produced. Certain local plants are also gathered and burnt, the belief being that the smoke will purify the air and protect the animal from disease and unwelcome spirits (Crate 2008: 120).

Protection rites form an important part of both human and animal life-cycles and, in many cultures, the act of marking or branding juvenile livestock is a magical device for ensuring the longevity of the animals and their human community, as well as being a symbol of identity and territoriality (as discussed in Chapter 5). Landais (2001: 472) explains how the Peuls of Niger mark animals to give them supernatural defence against death but also to promote fertility. The same reasons are frequently given to explain human tattooing and scarification; for instance Aufderheide and Rodrigez-Martin (1998: 49) highlight these blood-drawing practices as important maturation rituals, particularly for young men. It is probably no coincidence that the regions of the world traditionally associated with human skin modification – Siberia, Sub-Saharan Africa and the Americas – also have strong traditions of animal marking, some evidence for which is preserved archaeologically (see Figure 5.2).

Amongst Andean camelid herders, ear marking is an important rite of passage for llamas and alpacas, signifying their initiation to sexual maturity (Dransart 2002: 90–100). Today, ear-marking rites take place at the *Señalakuy* feast, an annual event during which herding communities come together to clip the ears of juvenile animals, mate older individuals and make sacrifices to the mountain divinities (Landais 2001: 472). According to Dransart (2002: 92–5) the focus of the festival is on animal fertility and lineage but the message is not lost on people, who use the occasion to consolidate and extend their human families. Indeed, Dransart (2002: 92) suggests that, originally, the camelid ear-marking festival would have paralleled the ear-piercing ceremony that was conducted for noble Inca boys, also as a maturation rite. In this way the human and animal populations of the ancient Andes would have passed through the major thresholds of their life-stages together.

A clear sign that Andean ear-clipping, and other forms of body marking, were previously associated with magic and ritual is provided by the fact that nearly all such traditions were curbed by the arrival of a new religion: Christianity. Dransart (2002: 92) mentions how Christian evangelization eradicated the practice of Inca ear-piercing. Similarly, in Europe, the spread of Christianity from the fourth century AD onwards gradually saw tattooing forbidden on religious grounds (Peoples and Bailey 2012: 349). To some extent the taboo against tattooing and scarification in ancient Christian, but also Roman and Greek, culture can be explained by the fact that is was considered bestial, a practice conducted by, or applied to, lesser beings: Barbarians, animals, criminals and slaves (Jones 1987). In turn, this reveals something of the worldview of these cultures and the separation they perceived between humans and animals. However, even where the connection between humans and animals is less intimate, animal marking may still have a ritual significance, be it to protect livestock from theft or to encourage lost animals to return home (Landais 2001: 475).

Rites of Passage (Movement)

Here I consider the word 'passage' in terms of movement and mobility, fundamental characteristics of life but also traits that have been the focus of many animal-based rituals throughout time and space. For instance, early Christian literature details a number of Anglo-Saxon charms that were centred on animal movement and, in particular, the retrieving of lost or stolen livestock (Olsan 1999). One Old English manuscript (Corpus Christi College 41) provides no fewer than four charms for recovering missing animals. In all cases saints' names are invoked and the practitioner is advised, in the case of a horse, to sing the charm over the animal's fetters or bridle. For lost cattle, the instructions are more complex with owners being instructed to drip wax from three lit candles in the hoof tracks of the missing animals whilst singing incantations such as 'Garmund, God's thane, find these cattle and fetch these cattle and have these cattle and hold these cattle and bring these cattle home' (Corpus Christi College MS 41, p. 206, cited in Olsan 1999: 411).

Such texts remind us that animals in the past were dynamic beings, sufficiently independent of mind and body that they could, and did, wander off on a regular basis: Olsan (1999: 409) suggests that the livestock charm must have been well known and possibly commonly used. The movement of livestock was clearly a concern to their owners; however, animal movements and behaviour could have far greater significance, with the capacity to dictate human actions.

Historical evidence from across the ancient Near East, Greece, Eturia, Roman and early Medieval Europe indicates that the movement and behaviour of animals was widely considered to be prophetic, carrying messages from the divine to the mortal world (Beal 2002; Johnston 2009). As was seen in the previous chapter (page 103), many cities – e.g. Rome and Thebes – were founded in either locations or formations indicated by the roaming of animals. Birds were deemed particularly prophetic, most probably due to their ability to negotiate the heavens in flight. Hittite texts preserve detailed information about bird oracles, recording how they asked questions of the deity, and received their answers by the way in which birds entered or exited a particular field of view and their behaviour therein – did they perch, turn, call, urinate or fight (Beal 2002: 65–8). Beal (2002: 70) notes that of all the Hittite oracles, bird diviners were the only ones where the name of the oracle is specified within the text, suggesting that this form of divination was highly specialized and regarded. By the time of Hesiod (*c.* 750–650 BC) however, lowly Greek farmers were instructed to 'judge the bird signs' before leading a new wife home (Johnston 2009: 129). The Roman author Plutarch (*c.* AD 46–120) reiterates the suggestion that the gods influence the behaviour of birds in order to advise humans about their actions (Bonnechere 2007: 11). To capitalize on this advice, the Romans kept flocks of sacred chickens, whom they consulted on the eve of important events: the chickens were fed and if they ate greedily it was considered a good omen (Johnston 2009: 130). But the omens were not always good. Shortly before the first Punic war, the Roman consul P. Claudius tried to induce the chickens to drink, because they had refused to eat, and inadvertently drowned them – needless to say, the Romans lost the war (Gilhus 2006: 26).

In general animals were favoured for divination because it was believed that they did not understand the message they were carrying, and therefore delivered it truthfully. The exceptions to this rule were the corvids – ravens and crows – intelligent creatures that were, according to Pliny (*Natural History*, 10.33) the only birds able to comprehend the messages they relayed (Gilhus 2006: 27). Corvids were particularly significant within Celtic, Roman and Germanic mythology, perhaps best exemplified by the Norse god Odin's raven companions, Hugin and Munin (thought and memory). Hugin and Munin acted as Odin's informers and envoys, flying out each morning and reporting back each evening on the events of the day (Serjeantson and Morris 2011: 100). They figure large in Germanic iconography (Kulakov and Markovets 2004) but similar representations are found in Celtic and Roman art, suggesting that the association between corvids, gods and prophecy was more widespread (Aldhouse-Green 2004; Green 1992; Serjeantson and Morris 2011). There is some historical evidence that ornithomancy endured into the Christian period of early medieval Europe, as it is frequently proscribed by Anglo-Saxon laws, which implies that the practice was occurring (Hinton 2005: 70). Archaeologically, fortune telling by the observation of birds is clearly difficult to detect; however, there is one example from Anglo-Saxon England that tempts the conclusion of ornithomancy. At the seventh-century cemetery of Lechlade in Gloucestershire, a young man was buried with both a bell (an object believed to have amuletic properties) but also a crow. The fact that this burial was placed at the outskirts of the cemetery, where 'deviant' burials are found in higher concentrations (Reynolds 2009: 201), may suggest that the community intentionally kept this person with fortune-telling powers at a distance (Hinton 2005: 70). In this instance, the crow clearly died alongside its owner, probably dispatched deliberately for the occasion of burial; however, animals need not die in order to be involved in funerary rituals, to which we now turn.

Funerary Rites

In Indonesia, live chickens play an important role in the Hindu cremation ceremony: their attendance during the rituals is thought to absorb evil spirits, preventing them from entering the bodies of the human mourners (Squire 2012: 80). After the ceremony, the chickens are released unharmed; so if similar practices occurred in the past we would know nothing of them, unless they were documented. For instance the Roman authors Cassius Dio and Herodian, both writing in the third century AD, record that birds were central to the funerals of Roman Emperors. They describe how imperial funerary rites culminated with the simultaneous lighting of the cremation pyre and release of an eagle, the latter symbolizing the ascent of the Emperor's soul to heaven (Arce 2010: 317–19). But animals can play a more direct role in human funerary practices, particularly when it comes to excarnation.

Amongst both the Parsis of India and the Buddhists of Tibet and Mongolia, the traditional funerary rite is to expose cadavers, with vultures and other carrion birds being encouraged to devour the corpses (Jacobi 2003: 814–5; Veraina 2013). The Parsis believe that dead bodies are the epitome of uncleanliness and that to bury, cremate or consign them to water would pollute the elements (earth, air or water respectively); so they seclude the dead within monumental open-topped excarnation towers, *Dokhmas*. In India, Dokhmas began to be constructed between the tenth to twelfth centuries AD, with earlier (third to seventh century AD) exposure platforms identified in Iran, the religion's homeland (Huff 2004; Nanji and Dhalla 2008; Veraina 2013: 76). For the Tibetan and Mongolian Buddhists, excarnation is motivated by a different set of beliefs; rather than seeing the dead as polluted, cadavers are viewed more as empty shells that require recycling in order to facilitate reincarnation. Here the vultures are seen as *Dakinis*, a psychopomp spirit, responsible for carrying the human soul to the sky, where it waits to be rehoused.

The role of birds as psychopomps has been proposed by Serjeantson and Morris (2011: 101) as one possible explanation for the large number of corvid remains recovered from the Iron Age hill fort of Danebury (Hampshire, England). At this site, approximately 1,200 bird bones were recovered, of which a staggering 80 per cent were corvids (70 per cent ravens and 10 per cent rook/crow). This is an unusually high frequency of corvids and their tentative connection with funerary rites is made stronger by the fact that disarticulated human remains were recovered in high frequencies at Danebury; a recent detailed study concluded that excarnation was practised at this site (Tracey 2012). Whilst the scavenging actions of corvids would leave little trace on skeletal remains, direct evidence for excarnation is clearly observable where canids are involved: gnawing marks are frequently found on Iron Age human remains, indicating that bodies were left exposed and accessible for local carnivores (Carr and Knüsel 1997; Craig *et al.* 2005; Redfern 2008). Similar traditions are found today amongst the Turkana and Nandi of Africa, who put their bodies out so that the hyenas can eat them and release their soul to the afterlife (Schwabe 1994: 47–8).

The excarnation practices of the Iron Age, and many modern zoocentric societies, compare dramatically with the funerary rites of the Romans, Christians, Jews and Moslems, all of whom valued proper burials. In stark contrast to the role of animals as sacred psychopomps, the Romans believed that souls were unlikely to reach the afterlife if the body was left exposed (Toynbee 1971: 43) and, for the Jews, to be scavenged by dogs was essentially to be cursed by God (Green 1997: 607). Again, these differences in death rites reflect differences in attitudes to the natural world (see Chapter 3) and highlight the importance of animals in the negotiation of belief systems and daily practice. It is these variations in worldview that determine whether animals will live or die as part of the funerary tradition. Clearly in the case of excarnation, it is a necessity that the animals are alive (and hungry) but it is equally clear that animals were often killed as part of human death rites.

Dead Animals in Rituals of Human Life

The most common zooarchaeological studies of animal rituals revolve around the concept of sacrifice, particularly where animal remains are found in obviously ceremonial situations, such as human graves or sanctuary/temple contexts. As these types of deposits are well considered in existing literature, I will not examine them in detail here, beyond offering two points of interpretation and terminology. The first is with regard to 'animals in human burials': how do we know that this statement, which gives primacy to the human, is correct? Given that many societies view humans and animals as equals, it is important to question, as Cross (2011) does 'who is buried with whom?' We must be open to the possibility that the human could have been sacrificed to accompany the animal.

Second, with regard to sacrifice, there is a tendency within zooarchaeological literature to consider issues of sacrifice only where animal skeletons are complete ABGs (Associated Bone Groups) and deliberately interred. However, in the majority of societies that lack commercial systems of meat redistribution, *all* animal slaughter is seen as sacrifice. For instance, in ancient Greek the word *mageiros* meant priest, but also butcher and cook (Symons 2002: 442). Most temples had kitchens where meat could be roasted and shared with the gods. Furthermore, the word for the knives used in the process, the *machaira*, are etymologically derived from *mageiros*, serving as a reminder that ritual, economics and material culture cannot be separated (Tsoukala 2009). With this in mind, we should expect that most of the animals that we recover from non-urban cultures are likely to represent sacrificed animals, even if we find only single fragmentary bones rather than complete skeletons. And we must also expect that, for these cultures, the moment of animal killing was not, as it is today, part of a production line but rather an emotive and sensory experience (e.g. Overton and Hamilakis 2013: 27–8).

Some insight into the experience can be gained from the anthropological literature, which shows that, across cultures, there are both differences but also overarching similarities in the approach to sacrifice.

Animal Sacrifice

Regardless of culture, few people actually enjoy killing animals. In the West we manage to ignore the realities of meat production and push the task of slaughter behind closed doors and into the hands of a small number of individuals, who become immune to the process thanks to its mechanical nature (Burt 2006; Wilkie 2010: 147–71). Psychologically this is assisted by a worldview that places humans at the top of the Chain of Being: animals are here for us to exploit. From a wider geographical and temporal perspective, it is probably safe to say that our culture is the weird one. Throughout time and space, most people have perceived animal killing to be an unpleasant experience, and where individuals have blood on their hands it has incurred feelings of guilt that need to be alleviated. Cue the soothing power of rituals and the supernatural.

Amongst the Sakha of Siberia, the killing of cattle is a sombre affair and not a regular occurrence, since most animals are considered part of the family and kept to an old age. Crate (2008) provides full details of the procedure, which begins with the animal being talked to by its owners: they rub its head, tell it that its life has been fulfilled but that its 'footprints are drying', meaning that others are now coming in its path. With great sorrow, the women withdraw from the scene, leaving the men to position the animal so that it faces the lake (from which cattle are believed to have been born) so that its soul might return whence it came. The animal is then knocked out with a blow to the head, the chest cavity is opened and the heart crushed – a practice that is believed to deliver the quickest death with the least suffering. An important point is that no blood is spilt as this would be considered to be a disrespectful waste of vital life force. Once the animal is dead, the women return to divide the carcass, and portions of each cut are fed to the spirits via the fire (Crate 2008).

The Suri of Africa show similar reluctance to kill their cattle, with sacrifice happening only on occasions of ceremony: for instance weddings, deaths and rain-making rites. They also share the belief that, in order to show respect, the animal must be killed without any blood loss. In this case, however, the blood taboo results in practices that, from the outside, appear very brutal indeed. Abbink (2003) describes one of the main forms of slaughter whereby cattle are led into a circle of people who bludgeon it with a ceremonial blunt pole, a process that can last more than twenty minutes before the animal dies.

This violent method of sacrifice is not an isolated example: Frazer (1912: 180–203) provides numerous case-studies of cultures who ritually slaughter their most sacred animals. The most detailed example is that which Frazer (1912: 182–5) provides about the Aino of the Japanese island of Yezo, for whom the bear is the main quarry, hunted as a source of protein and fur, but it is also the focus of religious activities. Frazer recounts the Aino tradition of catching bear cubs, which are brought to the village and treated with the utmost respect, suckled by the women and raised as a member of the family. When a bear becomes too big to handle, it is moved to a cage, where it continues to be cherished and well-kept. After two or three years, the bear is sacrificed. It is tethered and led out into the community, who, after reassuring the animal how much they love it, shoot large numbers of blunt arrows at it. As the bear becomes angry and struggles, it is gagged and strangled, its neck being crushed between two poles that the community work together to compress, again ensuring that no blood is spilt. Once dead, the animal is skinned, decapitated and its head is placed on display outside the bear's family home. Prayers are said to the animal, and then its flesh is boiled, the meat distributed to the entire community and the bones kept as sacred objects (see also Simoons and Baldwin 1982: 424–8).

It is possible to come closer to the psychology of the Aino's practices by considering the different elements in turn. It is easy to understand why a community that regularly hunts bears may wish to repay their spiritual debt by lavishing attention on a particular individual – the care they show their special house bear helping to absolve them of the guilt they acquire through hunting activities (an issue to which we shall return in Chapter 7, page 140). Why the community should chose to fire blunt arrows at the bear prior to killing it is more of a conundrum but I would suggest that it is a group mechanism for distancing the bear from the human community – the bear's anger returns it to a state of wilderness, rendering it a legitimate target for killing.

The examples of the Aino bear-killing may have some relevance for understanding the 'ritually-hunted' domestic cattle and pigs from Neolithic Durrington Walls (discussed on page 52). In a period when humans and animals must have been forming close relationships, the killing of a non-human member of the group would surely have been difficult psychologically. By taking well-known, and even well-loved, domestic animals into enclosures and hunting them communally, bonds and ties would have been severed as the livestock were driven from the human world and converted to the status of 'animal', through the act of violent abuse.

In the case of the Durrington Walls animals, blood must have been shed, as the arrows pierced the animals' skin becoming embedded in their bones. This would seem to hint at a different form of sacrifice to those considered above. For whilst many ritual killings aim to avoid blood loss, certain rites require it. Today, one of the most contentious blood sacrifice ceremonies is the Hindu Gadhimai festival, which takes place every five years in southern Nepal. At this festival, which lasts one week, approximately half a million animals (primarily buffalo, goats and chickens) are beheaded with ritual cleavers as a sacrifice to the God Gadhimai in request for support and fortune (Brockman 2011: 179).

Propitiatory rituals are commonly recorded within modern anthropological literature. Wessing and Jordaan (1997) reviewed the evidence for construction sacrifices in Southeast Asia, where animals are sacrificed in advance of building projects. They argued that these rites have two principle aims. The first is one of appeasement: that any form of soil-breaking for building purposes requires cooperation from the spirits who own the land, so a sacrifice is offered by way of gaining permission. The second aim concerns protection: buildings represent considerable investment and, by making a sacrifice to the spirits, there is some guarantee of protection, a form of supernatural insurance policy.

Within the archaeological record, there are many ABGs that would appear to fit this form of construction sacrifice. For instance Popkin (2010) found compelling evidence at Kilise Tepe in Turkey. Here, a complete sheep skeleton (minus the astragali – which are discussed further on page 128) was recovered in a pit that was located next to an altar stone in a ceremonial building. The animal was apparently interred towards the beginning of the building's life and examination of the animal's axis vertebra highlighted clear cut marks corresponding to where the jugular vein was repeatedly slashed (Figure 6.1). The possibility that the Kilise Tepe animal represents a construction or foundation sacrifice is perhaps

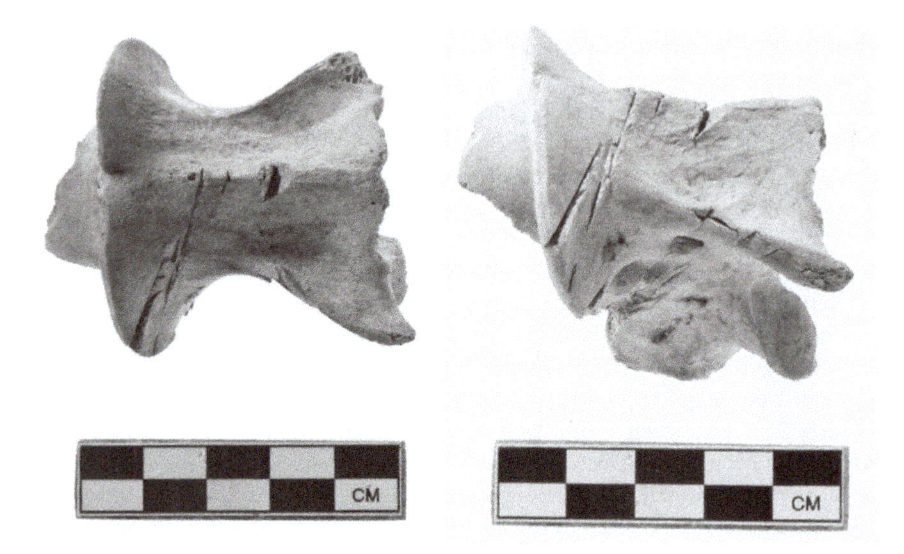

Figure 6.1 Heavily butchered axis vertebrae dorsal (left) and right side (right) views.
Source: Bob Miller and Peter Popkin.

strengthened by the many contemporary Hittite texts that explain how pits should be dug and blood spilt down into them from animals sacrificed as offerings to the chthonic deities (Collins 2002; Ünal 1988: 103).

Historical evidence reveals something of the structure and methods of animal sacrifice, with very detailed information being available for the Hittite, Greek, Etruscan and Roman procedures (Gilhus 2006; Mirecki and Meyer 2002). Taking an overview of the historical and anthropological literature, it is possible to identify several general stages of the sacrificial process, which may be outlined as follows:

1. Selection Different cultures choose different kinds of animals (male or female, young or old, wild or domestic, particular coat/plumage colours) for specific rituals. However, it is an almost universal truth that the animal to be sacrificed must be in prime condition, as to choose feeble or sickly animals would be a slight to the gods. This point completely overturns the 'economically rational' interpretations of the 1980s that envisaged sacrificial animals to be unwanted or valueless individuals (e.g. Grant 1984).

2. Preparation During this stage, introductory rituals (prayers, blessings, expressions of grief) are made and the animal may be decorated with garlands – a practice common to the Graeco-Romans (Gilhus 2006: 117) as well as the modern day Sakha and Suri (Abbink 2003; Crate 2008). The animal may be taken on a long or short procession; the former is depicted in much ancient Mediterranean art (Gilhus 2006: 117), the latter case is exemplified by the Aino who walk their house bear three times round its master's home before it is led to its death (Frazer 1912:

192). Importantly, in nearly all cases, the animal must be a willing participant in the sacrifice and understand what it happening – if it is perceived to be resistant, this would be a bad omen. Evidence for such beliefs can be found in ancient texts (Gilhus 2006: 119), and modern anthropological literature (e.g. see Brockman 2011: 179 for the Gadhimai festival, where it is believed that it is a bad omen if the animal makes a sound during the sacrifice, and Crate 2008 for the Sakha).

3. *Liminality* Many cultures have a stage whereby the victim is transferred from the human to the divine sphere: in the case of Roman procedures, animals were smeared with salt and flour (immolation) but this stage of transference could also account for the distancing procedures utilized by the Aino in their bear sacrifices.

4. *Death* This may or may not involve the spilling of blood, depending upon the purpose of the sacrifice. Wherever possible, zooarchaeologists should strive to determine the method of death as it has the potential to inform upon both the motivation for the sacrifice but also human attitudes to animals. The more 'respectful' and zoocentric rituals, often relating to community cohesion, are less likely to spill blood, so archaeologically we might expect evidence for blows to the head, crushing/compression fractures or simply no evidence at all. For blood sacrifices, which tend to be more anthro-centric and linked to individual or community fortune, we should perhaps look for evidence of decapitation or slitting of the jugular (e.g. Figure 6.1). Some care must be taken with the interpretation of archaeological material because heads are often severed after death for purposes of respectful display, as in the case of the Aino bear sacrifice (Frazer 1912: 182–5). One archaeological example that deserves comment in this regard is the Viking settlement of Hofstaðir in Iceland. At this site large numbers of cattle skulls, with clear evidence of decapitation, were recovered from within a monumental hall structure (Lucas and McGovern 2007). The remains were interpreted as evidence for 'bloody slaughter' and ritual beheading associated with human feasting and political empowerment. However, the skulls also demonstrated evidence of 'depressed fracture of the frontals caused by a heavy and immediately fatal crushing blow between the eyes' (Lucas and McGovern 2007: 10). If the cattle were killed by a blow to the forehead, rather than beheading, this indicates a very particular human–animal relationship more akin to the familial bond seen amongst the Sakha (Crate 2008). The fact that the heads were then taken and displayed after death would also seem to point towards a relationship of affection and memory, since many cultures perceive the head as the repository for the soul (Frazer 1912: 182–5; Lokuruka 2006). This indicates a scenario of greater complexity than the simple slaughter of anonymous cattle for commensal politics and the display of their skulls as an expression of human social status.

5. *Haruspication* The final stage, common to all sacrifices that I have examined in the ancient (e.g. Brown 2006; Collins 2008; Gilhus 2006) and anthropological (e.g. Abbink 1993) literature is examination of the entrails. It is a practice about which the zooarchaeological literature is almost entirely silent, presumably because it

involves soft tissues. However, haruspication is, almost universally, a culturally important act and one deserving further consideration. Russell (2012: 127–30) has provided a solid overview of haruspication, bringing together the historical, anthropological and artefactual evidence for the practice: this work should be consulted for further information. In summary, at the point of death, the animal becomes a medium for communication with the gods, who will leave messages in the animal's organs and intestines. The tissues are usually examined by specialists – trained diviners (see Brown 2006; Collins 2008), shamans or village elders (Abbink 1993) – who decipher the messages and convey them to the onlookers. The important generalization here is that, in order for the message to be of relevance, the animal must be linked to community: if an individual seeks information from the divine, it can only be mediated through an animal belonging to that individual. By example, in his study of the Suri, Abbink (1993: 709) questioned a man who was sacrificing one of his goats in order to find out the fate of his ill wife. Abbink asked whether it would be possible to use a wild hartebeest or a chicken instead of a goat and the reply came back with incredulity: 'of course not . . . what could they possibly tell?' The same principles appear to have applied in the past, as within both Greek and Roman practices sacrifice was largely restricted to livestock rather than animals from beyond the domestic sphere (Gilhus 2006: 115 – although see Larson, in press, for the sacrifice of wild animals to the Greek goddess Artemis).

From the information given above it becomes difficult to draw any kind of definitive lines that separate animal killing from religion, from magic, or from medicine: all are combined and the animals are not passive ingredients in the mix. With the exception of highly urbanized societies, we must expect that all of these social elements are reflected in the animal bones that we study. Even where animal remains appear to be innocuous domestic food waste they are likely vestiges of complex interactions and belief systems; we are just very bad at identifying and understanding them. Our interpretative inabilities extend beyond the rituals of animal sacrifice, as zooarchaeologists also have a poor track-record of elucidating the role of animal products and body parts in magic and, in particular, medicine.

Animal Magic and Medicine

As ever, it is the obviously unusual deposits – modified bones and teeth or collections of particular skeletal elements – that have attracted the most attention from zooarchaeologists in search of magic and medicine. Animal remains found in association with human burials are particularly conspicuous within the literature, such as 'amuletic' tooth, bone or shell pendants, which have been recovered from sites across Prehistoric Europe and the Near East (Horwitz and Goring-Morris 2004; Mannermaa 2008; Vanhaeren and d'Errico 2005), the ancient and historical Americas (Mensforth 2007) and beyond (e.g. Handler 1997).

Today the wearing of animal-derived pendants is widespread within non-Western societies but may occur for a variety of reasons, some operating simultaneously. For instance, pendants obtained from hunted animals may be worn, on one level, as an indication of social status, perhaps demonstrating and symbolizing a hunter's accomplishments; however, on another level, they may provide spiritual sustenance, giving the hunter ownership over the animal's power (Mensforth 2007). Certainly, the KhoeSān of Africa give ostrich shell necklaces to weak or sick children to provide them with the ostrich's potency (Low 2011: 302–3). By contrast to this healing role, in East Java tiger remains are worn for purposes of aggression: it is believed that tiger's bones, teeth and blood contain the properties of the animal that will be transferred to those who possess the remains, allowing men to become supernaturally strong and invincible in fights or warfare (Wessing 1995). The potency of animals can, however, be obtained without their physical remains; according to Wessing (1995: 201) proximity to a tiger image is sufficient to benefit from the animal's properties. The possibility that archaeological cultures similarly valued animal iconography for reasons of empowerment is exemplified by finds of weaponry and armour bearing the images of wild and ferocious animals: e.g. the boar crest warrior helmets of Bronze Age Greece (Hamilakis 2003) or the carnivorous animals that adorn martial equipment of early medieval Europe (Dickinson 2005; Pluskowski 2010). There is also evidence that images were used for their apotropaic properties to protect the vulnerable. Crummy (2010) has identified a consistent pattern of grave goods in infant burials from Roman Britain, where both cremations and inhumations contain representations of bears. When Crummy (2010) viewed these finds in their wider cultural context it become clear that the bear motif – an animal associated with Artemis, the goddess of childbirth and child-rearing – was deliberately selected to ensure that the child did not enter the underworld alone or unprotected.

The physical and cultural context in which animal remains are found, and their association with other materials, certainly helps the process of interpretation. Indeed, Greaves (2012: 191) stresses that understanding the context is vital, although in itself, it may not be sufficient to draw conclusions about a given deposit. Greaves's work relates particularly to elucidating the role of the astragalus within ancient Greek divination practices. This skeletal element, a small but high-density bone with a pleasing shape and feel, has been the focus of numerous zooarchaeological studies concerning magic because, across cultures past and present, it has been incorporated into cultic activities (Affanni 2008; Dandoy 2006; Gilmour 1997; Koerper and Whitney-Desautels 1999). However, the astragalus is also used in more secular spheres, such as in gaming and gambling, where the knuckle-bones are used as counters, so interpretations are not always straightforward (see Russell 2012: 133–7). Greaves has proposed three criteria to identify the cultic function of astragali: the first is context – if they are found in sanctuaries it is probable they have a magico-religious significance; the second is modification – some archaeological specimens demonstrate shaving, smoothing or may even have had lead insertions to change the way they fall when thrown; the third is inscription – many sites in the Mediterranean have yielded astragali incised with the names of gods and heroes (Greaves 2012: 192).

Inscribed astragali have also been found in northern Europe, with one incised example recovered from a cache of thirty deposited in a fifth-century AD Anglo-Saxon burial urn in Norfolk, England (Myres and Green 1973). Of the astragali, twenty-nine were identified as sheep, the last being a single roe deer astragalus, which was also the one inscribed, with runes, 'ræȝhæn'. Initial analysis of this find concluded that the deposit was evidence of divination, suggesting that the runes were cognate with Old Norse *regin*, 'gods'; however, this interpretation was later challenged on linguistic grounds and it is now accepted that 'ræȝ hæn' is a cognate of Old English *raha*, 'roe deer', a more fitting interpretation given the source of the bone (Page 1995: 123). Rather than representing divination, the astragali deposit is now assumed to represent a form of board game (Barnes 2012: 42). However, gaming and divination are often closely linked, with many people believing that the supernatural is responsible for the results of both (Greaves 2012: 192; Koerper and Whitney-Desautels 1999: 74; Russell 2012: 136).

More convincing examples of the link between astragali and divination are found in Anatolia, where astragalomancy is recorded in Hittite texts dating to the thirteenth and fourteenth century BC (Greaves 2012: 191). Returning to the site of Kilise Tepe, which provided good evidence of a ritual blood sacrifice (see page 123), it is interesting to note that both the animal's astragali has been deliberately removed from the sacrificed animal. Furthermore, caches of astragali were recovered around the temple site (Popkin 2010). Similar astragali deposits have been noted from Iron Age levels at Tell Afis, Syria, and their association with wall foundations led Affanni (2008) to interpret the deposit as a construction rite. It seems highly likely that these skeletal elements, as with those from Kilise Tepe, had been extracted from sacrificed animals. Affanni (2008: 86) argues that astragalomancy was intimately linked to the sacrifice and the entrails reading; thus the astragalus represented a more sustainable and repeatable form of divination than the single event of soft-tissue haruspication. The fact that several of the Tell Afis astragali demonstrated signs of polish (indicating repeated handling) whilst others had been smoothed or burnt suggests, as Affanni (2008) concludes, that the bones of sacrificed animals played an important role in the magic of the ancient Near East. There is considerable evidence that other skeletal elements, in particular shoulder-blades, were also utilized in divination practices across the Near East and Mediterranean, with many incised or burnt scapulae being recovered from temples and sanctuaries (e.g. Marom *et al.* 2006; Reese 2002; Russell 2012).

There is a temptation to see these sacrifice-divination practices as traits of Pagan cosmologies, especially since Christian authors, notably Isidore of Seville and Augustine of Hippo, document strong opposition to both animal sacrifice and divination (Gilhus 2006: 26, 166; Jolly *et al.* 2002). However, there is increasing evidence that in geographical areas of religious syncretism, Christian dogma was overlooked, such as in medieval Spain where Christians adopted Muslim techniques of scapulimancy (Burnett 1997: 100–35; Green 2006: 29). Indeed, the Bodleian Library in Oxford holds one thirteenth-century document (Canon Misc 396, fol. 112r) containing a very detailed map of a scapula, illustrating and describing the significance of major landmarks, to guide the diviner in their reading of the shoulder-blade. We should, therefore, not assume that all traces of animal magic died out with the coming of Christianity: Burnett (1997) and Jolly *et al.* (2002) demonstrate just how widespread divination and other magical practices were in medieval Europe, with notable increases occurring in the late fourteenth and fifteenth centuries (Jolly *et al.* 2002: 23) most probably in response to the trauma of the Black Death (Gilchrist 2008). With this in mind, we must be prepared to look for, and find, evidence for such practices in the archaeological record; although, as Gazin-Schwartz (2001) has highlighted, magico-religious practices were likely so imbedded in daily practice that they become invisible due to their ubiquity. The same explanation may account for the apparent absence of animal-derived medicine, or 'zootherapy', in the archaeological record.

Russell (2012: 392–4) is one of few zooarchaeologists who has explicitly set out to examine evidence for zootherapy. She states that archaeological examples of animal-derived medicines are difficult to find, although admits that few people have tried (Russell 2012: 393); I suspect this is the real reason for the apparent absence of evidence. Anthropological research makes clear that zootherapy is everywhere in non-Western societies, the World Health Organization (WHO, 1993) estimating that about 80 per cent of the world's population – six billion people – rely primarily on animal and plant-based medicines (e.g. Alves and Rosa 2005; Alves *et al.* 2013; Costa-Neto 2005; Lev 2003; Morris 1998: 214–29). Likewise, ancient texts attest to the ubiquity of zootherapy: Lev (2003: 108) concludes that medical treatises on animals emerge with the origins of writing itself, highlighting the centrality and longevity of zootherapy within human society. Given the widespread use of animal medicines in societies past and present, it is inconceivable that the remnants of such practices are not preserved in abundance within the archaeological record. The difficulty is identifying them. There are two main reasons why this is so. The first is related to our own worldview: for us, medicine and medical practices are distinct practices, set apart from other forms of food/drink or behaviour respectively, but the same is not true of other cultures in time and space. One needs look no further than humoural principles (see pages 17–19) to realize that, for most societies, everything that people consume, through any of their senses (taste, touch, sight or sound), has the potential to influence an individual's well-being. In this way, we must expect that all animal-derived foods and objects, however functional, had the potential to be used medicinally.

A good example of this is the ostrich egg: from the earliest Babylonian and Assyrian texts they are recorded as having medicinal qualities (Finet 1982: 75) and were widely traded around the Mediterranean from the Bronze Age to the medieval period (Green 2006: 31–2). Often they were worked into cups or rhyta, such as the example from Kish, Mesopotamia, which is dated *c.* 3000 BC. Based on humoural principles and the belief that the temperaments of different organisms could be transferred through the senses, it seems probable that the act of drinking from an ostrich egg would be medically beneficial. If this case can be made tentatively for the ostrich egg, it can be made with certainty for the unicorn goblets of late medieval and post-medieval Europe (Schoenberger 1951). Several examples of elaborate drinking vessels made from 'unicorn' (*Monodon monoceros*) horn survive within museum collections, a mere fraction of the examples recorded in household accounts and other documentary sources dating back to the fourth-century Greek author Ctesias (Schoenberger 1951). The demand for these goblets derived from the belief that any liquids drunk from such an object would either cure or prevent poisoning or disease (Schoenberger 1951: 284). Unicorn horn could also be taken in powdered form; indeed many of the unicorn goblets are known to have been ground down for this purpose, which leads me to the second reason why animal-based medicines may not be apparent archaeologically: because the very act of converting them to medicine is destructive.

Today, zootherapy relies upon powdered body parts (Lev 2003) and animal products. For instance the KhoeSān use powdered ostrich shell as an important medicine against fever, diarrhoea and stomach pain (Low 2011). Similarly Pliny's treatise on the medical properties of animals frequently calls for their remains to be ashed and powdered. As with modern Chinese medicine, Pliny rates deer antler highly, suggesting that it can be powdered to cure ills from tooth ache to epilepsy (*Natural History* Book XXVII, trans. Jones 1963). Elsewhere I suggested that such practices could account for some of the shaving marks seen on fallow deer antler from Roman northern Europe (Sykes 2010c). This possibility seems strengthened when the skeletal representation data for fallow deer are examined (Table 4.1). In Iron Age and Roman Britain, shed fallow deer antler is the most common body part and the possibility that these skeletal elements were imported from the Mediterranean to Britain for medicinal purposes has recently been supported by multi-element isotope analysis (Madgwick *et al.* 2013). Recent oxygen isotope analysis has confirmed that, in all probability, these specimens originated from the Mediterranean (Miller *et al.* submitted).

It may seem bizarre that the population of Roman Britain should have taken the trouble to import fallow deer antler when the antlers of native deer were readily available; however, in much the same way that geographical distance is frequently associated with supernatural distance (Helms 1993), so healing power is often deemed to increase with cultural distance (Rekdal 1999: 473). For this reason, zootherapy often preferentially targets the remains of animals that are exotic or wild, coming as they do from beyond the realm of daily practice. It seems possible that this situation may account for an important discrepancy between the archaeological and historical evidence for the Roman period: namely that exotic animals – particularly elephants and big cats – are recorded in high frequencies in the documentary sources but their remains are almost entirely lacking in the zooarchaeological record (MacKinnon 2006), is it possible that their bones were destroyed in the manufacture of the medicines detailed by the likes of Pliny?

The only way we could confirm such a scenario would be to find remnants of bone that exhibit signs of having been powdered or ashed. At present, no such examples exist for the Roman period. For the Mesolithic period, however, there are some bone specimens that lend themselves to such an interpretation. Lage (2009) examined numerous wild mammal remains recovered from Maglemose and Ertebølle sites in Germany that demonstrated unusual and heavy erosion on their shaft, where the bone had been smoothed away by a process that left chatter marks on the bone surface. Experimental replication of the chatter marks was successful only when a flint blade was scraped across the bone surface. This resulted in the production of fine bone powder, which Lage (2009) suggested may have been used for medicine. I find this interpretation very convincing and would recommend that other zooarchaeologists are aware that this kind of bone modification may be direct evidence for medicine production: we just need to look for it.

Summary

Zooarchaeologists have always gone in search of the 'unusual' and the 'special' to hold up as evidence of animal-based rites. To be sure, ABGs can often be linked to animal sacrifices; however, the concentration on dead animals in obviously religious settings (e.g. temples or cemeteries) is restricting research on animal rituals to a mere fraction of their original occurrence. We are simply projecting our 'Church on a Sunday' compartmentalization of the sacred and profane back onto societies that saw no such division. Beyond our own 'Western' society, all aspects of daily life – economics, diet, religion, magic, medicine, the elements and landscape – are experienced and understood together.

We may expect that, for the majority of archaeological societies, all animal killing would have been sacrifice. But animals do not become 'ritual' only at the point of death, and we must try to consider how the living animals helped to co-create human culture, daily life, religion and decision making.

The role of animals in human well-being, be it psychological or physical, is a subject that is woefully under-researched in zooarchaeology and deserves far greater attention, especially given the prominence of zootherapy, animal magic and animal medicine within the anthropological and historical literature. Today, an important area of zootherapy is pet-keeping, which medical trials have demonstrated can be exceptionally beneficial to human well-being (Costa-Neto 2005). It is to the issue of companion animals that we now turn, as we consider the evidence for cultural attitudes to animals as friends, foe and partners.

Chapter 7

FRIENDS, CONFIDANTS AND LOVERS

Anyone who has a pet or lives with a 'companion animal' or works on a farm will know that animals are not all the same: dogs are not all the same, nor are all cats, or cows, horses, chickens or even bees. As individuals they vary in character and temperament, and human relationships with them are equally diverse and ever-changing, falling anywhere on a cyclical spectrum from the most intimate and loving to the most cruel and brutal. Whatever the form of the relationship, it always reflects human ideology, be it individual or group-based. As has been shown repeatedly by anthropologists (e.g. Knight 2005) there is every reason to try and identify the nature of interactions between humans and (living) animals, and unpick their meaning and cultural significance.

Sociologists, psychologists and historians have made great progress in charting long-term shifts in cultural attitudes to animals (Ritvo 1987; Ryder 2000; Serpell 1996; Thomas 1983), with detailed studies examining the rise and fall of pet-keeping cultures (Bodson 2000; Podberscek *et al.* 2000; Serpell 1996; Walker-Meikle 2012), as well as animal cruelty and bestiality, or 'zoophilia' (Beetz and Podberscek 2005; Beirne 1994; 2000; Dekkers 2000; Miletski 2005). Zooarchaeologists are increasingly making a contribution to the debate, particularly with regard to pet-keeping (Mackinnon 2010c; Salmi 2012; Thomas 2005b) and issues of animal-directed violence (Binois *et al.* 2013; O'Regan 2002; Teegan 2005); however, traditional zooarchaeological studies have been limited by ambiguous data.

Thomas (2005b: 96) sets out the types of animal bone evidence that might be examined by those in search of pets, but none of the sources mentioned are conclusive in themselves. For while we may expect that a pet may be given a formal burial, the deposition of complete carcasses in unusual settings may equally result from sacrifices linked to religion (White *et al.* 2001; see Chapter 6) or medicine (De Grossi Mazzorin and Minniti 2006), or the deliberate removal of animals considered pests or unclean (Wapnish and Hesse 1993). Pathology has the potential to inform on patterns of cruelty or care but such evidence needs careful consideration since lesions may have many possible aetiologies (Thomas 2005b: 95). The conformation of dogs, in particular the presence of miniature 'toy' breeds, is often held up as evidence for pets but how can we know whether we are looking at the remains of pampered lap dogs rather than working ratters? Similarly, the presence of exotic animals – monkeys, parrots or tortoises – are frequently labelled as pets; however, it is difficult to be certain whether they were considered and treated as cherished confidants or abused sideshow freaks: the bones alone do not reveal this (Albarella and Thomas 2002; Brisbane *et al.* 2007; Thomas 2010). Even if we were able to classify confidently a set of animal remains as having derived from a 'pet', this label does not adequately account for the likely complexity of the human–animal engagement: no relationship is that consistent and it is quite possible for individuals to be adored one day and assaulted, physically or sexually, the next.

Currently zooarchaeological investigations of these issues are, by necessity, so coarse that they cannot provide the refinement necessary to understand the nuances of human–animal relations. This can only be achieved by focusing on individual-level analyses so that we might gradually build a robust dataset from which wider cultural interpretations can be drawn. One of the most promising avenues in this regard is isotope analyses, as these provide the best opportunity to gain fine-resolution information at the level of the individual. Unfortunately, for such techniques to be worthwhile, we need to break with the intellectual rationale of traditional isotope analyses, where animals are studied almost exclusively as test-beds for establishing the parameters of isotope variation (Balasse 2002; Bocherens and Drucker 2003; Richards *et al.* 2003; Sponheimer *et al.* 2003), as background data to better understand 'what people ate' (e.g. Müldner and Richards 2005, 2007) or, in the case of dogs, as 'proxies' for human diet (Guiry 2012, 2013). Within archaeological science, animals are seldom considered as individuals in their own right (although see MacKinnon 2010c). I believe this situation needs to change and that, if it does, isotope analyses will have the potential to transform our understanding of human–animal relationships.

If we are to succeed in our examination of *relationships* it is important to recognize that these cannot be understood by examining only one of the parties: if we focus our attention purely on animals, we will gain only one half of the story. Essentially, human culture and human–animal relationships cannot be understood without considering all of the participants, and in this chapter I will examine how attitudes to animals inform on broader issues of social construction and identity definition. I will start by examining issues of animal cruelty and blood sports, in particular cockfighting, before moving on to consider zoophilia and varying attitudes towards it, finishing with a consideration of 'pet-keeping'. It will become clear that none of these categories is straightforward, often merging from one to the other, and all three forms of relationship – cruelty, bestiality and pet-keeping – are interlinked in ideological terms, with human attitudes towards each showing complementary trends.

Blood Sports: Loving Animals to Death

Within English historical studies, the sixteenth and seventeenth centuries are believed to represent the most sustained period of animal-directed violence witnessed in the human past, when blood sports such as animal baiting, hunting, coursing and cockfighting were the popular culture, detailed graphically by the literature and art of the time (Brownstein 1969; Kalof 2007; Ryder 2000: 42; Serpell 1996: 154–9; Thomas 1983). The other notorious episode of animal cruelty is that of the Roman period with its arena-based torture and slaughter of animals, in particular exotics, for public amusement (Auguet 1972; Kyle 1998). Both periods were moments of imperial expansion and aggression that resulted in the domination of indigenous populations and it may be no coincidence that in these periods of male-dominated military conquest, hunting and bloods sports – both of which are universally linked to aggression and masculinity (see Chapter 3) – became particularly prolific. Whilst links between imperialism, martiality and animal-directed violence are appropriate, the Roman and Renaissance periods stand out precisely because they are well documented; other cultures that were equally passionate about blood sports have slipped under the radar due to lack of historical documentation.

Recent research (Doherty 2013; Hodkinson 2013) has highlighted that cockfighting was far more widespread than has hitherto been recognized by either historians or zooarchaeologists, most likely being endemic throughout Europe since the Iron Age. Today the sport is almost entirely outlawed and considered to be one of the most brutal forms of animal abuse and exploitation. Whilst I do not wish to suggest otherwise, anthropological evidence indicates that the situation is not as clear-cut as it might first appear. As set out in Chapter 4, cockfighting societies seldom see their cockerels as fungible automatons. On the contrary, birds and their keepers tend to be closely connected and the fortunes of both parties rest on their ability to work together (Dundes 1994). Because cockfighting is frequently bound up with spiritual, social and financial concerns (it is usually linked to gambling), men are often obsessed with their roosters, treating them with great care and attention to ensure their wellbeing (Dundes 1994). In the archaeological record such interactions are difficult to elucidate; however recent research on an Avar period (seventh/eighth century) cemetery from Wien-Csokorgasse in Austria is providing some insights into past human–chicken relationships (Kroll 2013).

Many of the human burials from Wien-Csokorgasse contain rich grave goods, some clearly indicative of high social status, others less so but one of the most common offerings is the chicken. Traditional zooarchaeological investigations by Kroll (2013) demonstrated that the deposition of these animals was not random but gender-based: women were interred with hens and men with cockerels. Interestingly, Kroll (2013) noted that the higher the status of human males, the longer the cock-spur of the rooster. This is an important point because spur length is age related, and work by Doherty (2013) has demonstrated that cockerels exhibiting long spurs must have lived for months if not years; plenty of time for a close bond to develop. To test the hypothesis that close human–chicken relations existed at Wien-Csokorgasse, samples from the chickens were submitted for stable isotope analyses in the hope of comparing the results with existing data for the humans (Herold 2008). Figure 7.1 shows the preliminary results from the study

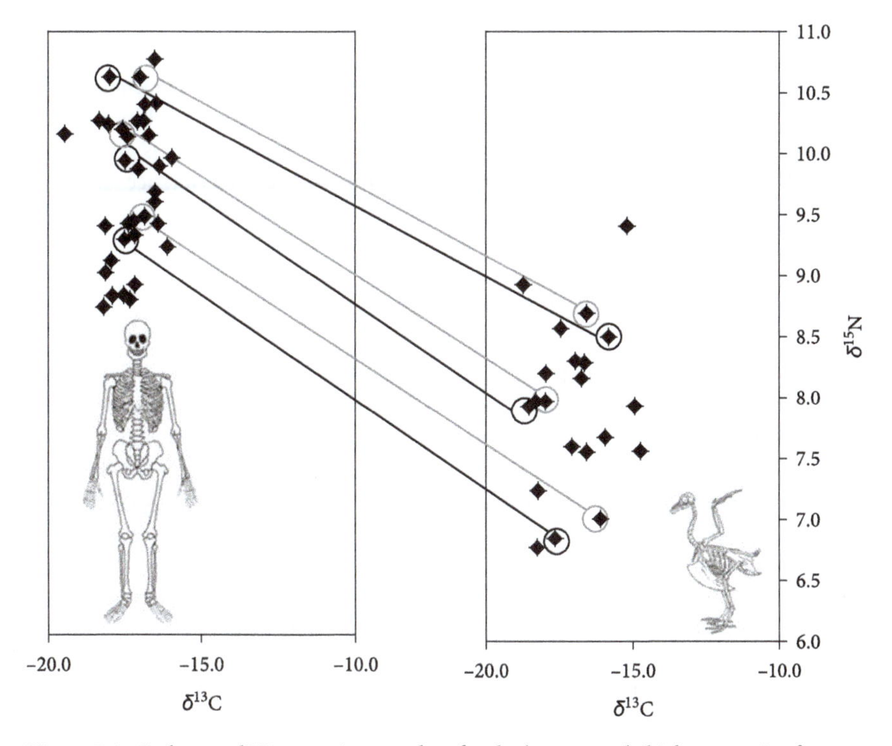

Figure 7.1 Carbon and Nitrogen isotope data for the human and chicken remains from the seventh-century cemetery at Wien-Csokorgasse, Vienna.

Source: Sykes.

and it can be seen that, when the results for the chickens are connected with those for the human individuals with whom they were interred, there is a direct correlation in the ranking of the $\delta^{15}N$ values: chickens with high nitrogen values were buried with high-ranking humans, whereas chickens with lower $\delta^{15}N$ values were interred with similarly low-ranking people. In effect, the humans and the chickens mirror each other in dietary terms, suggesting that relationships between the two were very close indeed: the people must have known and been commensal with the chickens with whom they were buried. One cockerel (that with the $\delta^{15}N$ value of 9.5) appears to have had a very rich diet, containing more protein than was consumed by many of the humans. This particular individual was interred with a male of 'unusually high status' and its cock-spur was very long (Kroll personal communication), indicating that it was an old and well-fed individual, perhaps a champion fighting bird.

Based on the anthropological evidence we might speculate about the kind of relationship that existed between this particular cockerel and its human companion. Dundes (1994) has highlighted the intimacies that often occur between men and their roosters. When cockerels are injured, men have often been known, across cultures, to take their bird's head in their mouths and suck them, or blow up the bird's vent in order to resurrect it. Such behaviours obviously attract accusations of zoophilia and the suggestion is not entirely unfounded. Indeed, many blood sports, in particular cockfighting and hunting, are overtly sexual and it is sometimes difficult to draw clear boundaries between animal-directed violence and love (Cartmill 1993; Luke 1998).

Indivisibility and Bestiality

Those who have researched bestiality, or 'zoophilia', agree that it is a practice common to every world culture throughout history, the only variables being the motivations behind the act and how it is perceived in social terms (Beetz and Podberscek 2005; Beirne 2000; Dekkers 2000; Miletski 2005; Serpell 1996). The typology of bestiality offered by Havelock Ellis (1900: 79), although a product of its time, is a good starting point:

> Three conditions have favoured the extreme prevalence of bestiality: 1) primitive conceptions of life which build up no great barrier between man and other animals; 2) extreme familiarity which exists between peasants and beasts; 3) folk-lore beliefs such as the efficacy of intercourse with animals to cure venereal disease.

More recent surveys would seem to support the validity of these three broad categories, in particular the medicinal benefits of intercourse with animals: a belief held by the ancient Egyptians, Greeks, and population of early medieval Europe, as well as recent historic and modern tribes of the world (Miletski 2005). However, most of the evidence for animal-directed sexuality suggests that it is frequently associated with rites of passage, magico-religious ceremonies or power negotiations, sometimes of vital importance for human life.

For instance, as demonstrated in Chapter 2, hunting is explicitly sexual and, across hunter-gatherer cultures – from the Amazon to Africa and Asia to the Circumpolar region – hunters aim to seduce animals into giving themselves up (see Willerslev 2007: 110 for numerous relevant sources). The implication must be that, without such psycho-sexual relations, there is a risk that no hunters would be successful. It is for this reason that most hunters avoid sexual contact with women prior to hunting trips, in order to preserve their sexual potency for the animals (e.g. Morris 1998: 102; Morris 2000: 85). Amongst these cultures, hunters are only on the peripheries of human–animal seduction, a more central role being played by the shamans who are required to transform into animals in order to negotiate with spirit masters. As part of this process, the Desana and Tapiapé shamans of the Amazon are said to copulate with peccaries (Erikson 2000: 12), whilst amongst the Sakha of Siberia the shaman gives birth to animals during periods of trance (Hollimon 2001: 126). Similar beliefs may account for some Palaeolithic and Mesolithic artworks, which show sexualized human–animal hybrids (Borić 2007).

Clearly many of these shamanic forms of bestiality are more psychological than actual, and this perhaps reflects the greater physical distance between humans and animals in hunter-gatherer societies (Chapter 2). Within farming communities, however, humans and animals live in closer proximity, providing greater opportunities for more intimate relationships to develop.

Based on wider historical and anthropological studies (Miletski 2005; Serpell 1996), it seems almost certain that bestiality occurred in the early farming societies of the Neolithic but evidence for such acts, and attitudes towards them, are difficult to discern. Iconographic evidence, in particular the frequent representation of human–animal hybrids (Borić 2007), may indicate a belief in human–animal transmigration, under which circumstances bestiality is frequently accepted (Miletski 2005: 10). Nevertheless, it is unlikely to have gone unnoticed and, if not tabooed, it would presumably have been ritualized. Russell and Düring (2006) describe an unusual double burial of an adult male and a female lamb at Neolithic Çatalhöyük and suggest that it may reflect a strong relationship between the two individuals during life. Given the rarity of such co-burials, this seems a likely interpretation. However, the fact that the burial area was later avoided, with subsequent interments being kept at a distance, hints at a certain level of unease: is it possible that the 'strong' relationship between the two was sexual and was not considered entirely appropriate?

The earliest documentary evidence for taboos over bestiality come from ancient Babylonian texts, such as the *Code of Hammurabi* (*c.*1900 BC), which states that the practice was punishable with death. Similar legislation is found in Hittite and Egyptian documents, although the latter indicates some magico-religious situations where bestiality was considered acceptable (Miletski 2005: 2–3). A level of tolerance should be expected from the ancient Egyptians, given that most of their gods were human–animal hybrids. The same is true for the ancient Greeks, many of whose divinities were the result of inter-species relationships. In common with many other cultures – both past and present – the Greeks traced their ancestry to primordial sexual encounters between women and animals (Serpell 1996: 33; see Crate 2008 for the Sakha, and Miletski 2005 for other anthropological evidence). Indeed, there is no evidence that bestiality was punished in ancient Greek society (Miletski 2005: 4). For the Romans, the situation was slightly different. Although bestiality was tolerated, as evidenced by Aelian's *De Natura Animalium* which provides plenty of account of inter-species sexual relationships, it was also mocked (Salisbury 1994: 86). Furthermore, Roman deities demonstrate a greater level of separation from animals: when depicted they are often shown in the *company* of animals – Mercury with cockerels, Diana with deer or bears and Epona with horses – rather than hybridized with them (see Figures 3.3a and 3.3b).

In areas of Europe outside the Roman Empire, attitudes to animal divinities and bestiality appear to have endured largely unchanged from the Iron Age into the early medieval period. As with the ancient Greeks, there is evidence that many groups traced their origins to bestial relationships: Irish myths recounting heroic liaisons with horse gods (see page 83), whilst early Danish kings claimed their lineage was founded by the offspring of a woman and a bear (Salisbury1994: 85). Scandinavia would appear to have something of a grand tradition of bestiality, dating back into Prehistory. In Sweden, for instance, sexual relations between men and animals (primarily horses and elk) are depicted in the famous Bronze Age (*c.* 1000 BC) rock paintings of Bohuslän, as well as those found at Lake Abo in Ångermanland and the stone circle at Sagaholm in Småland (Jennebert 2011: 181–2). Artistic representation of human–animal hybrids, although less overtly sexual, continue through Iron Age and into pre-Christian medieval period suggesting that the boundaries between humans and animals were perceived as blurred (Kristoffersen 2010). Jennbert (2011: 182) suggests that Scandinavian bestiality should be interpreted as a ritualized manifestation of power. Given the emphasis on sexual relationships with horses, her suggestion is consistent with the bestiality kingship rites of the Aśvamedha, which were seemingly practised in some form across India, the Mediterranean and into Europe, their spread coincident with the diffusion of the horse (Chapter 4).

In northern Europe, Early Germanic secular law makes no mention of bestiality, suggesting either that no one was doing it or, more probably given the wider evidence, that no one cared (Salisbury 1994: 87). Gradually, with the spread of Christianity, rumblings about the sinful nature of human–animal liaisons became increasingly louder and penalties for it increasingly severe. Opposition to bestiality can be traced to two fundamental points. The first is that, in common with all monotheistic religions, Christianity placed humans at the top of the Chain of Being, their supremacy over, and separation from, animals established by the almighty God. Bestiality corrupted this natural order, reducing humans to mere animals, one of the most heinous offences to the Christian mind (Gilhus 2006: 212). Second, it was believed that any union between humans and animals would result in the production of 'monstrous offspring' (Beirne 2000: 318), a clear attack on the human–animal hybrid deities of pagan religions from which the Christian Church sought to differentiate itself.

Salisbury (1994: 89–90) details how punishments for bestiality were fairly lenient in the earliest Christian penitentials of the seventh to eighth century, written whilst the new faith was trying to get a foothold in the Pagan north: a youth caught in the act with an animal would be required to do between forty and one-hundred days penance, similar to the penalty for the sin of masturbation. A few centuries on, however, bestiality had become equated with homosexuality which, according to the Biblical injunctions set out in Leviticus (20. 15–16), required the execution of all parties (Gade 1986: 127). Although it was often possible to commute capital punishment for cash payments, there is some archaeological evidence to suggest that humans and animals were occasionally executed for their crimes against nature. The most compelling examples come from the Anglo-Saxon execution cemeteries of Old Dairy Cottage and Stockbridge Down in Hampshire (England) where decapitated men were buried with animals (for full details see Reynolds 2009: 172). At Old Dairy Cottage, burial 565 contained a decapitated man who was buried with four neonatal lambs, animals that could, potentially, have been viewed as the 'monstrous offspring' begotten of an illicit union between man and sheep. At Stockbridge Down burial 37 was of a man who was interred with the head of a sheep, which Reynolds (2009: 172) interprets as potential evidence for bestiality given a near contemporary (*c.* 955 AD) charter bound for Chalke in Wiltshire, which records the location where a man was executed *for þan buccan*, 'because of the goat'. A second execution burial at Stockbridge Down (number 19) contained a decapitated man and a decapitated dog; the fact that both were treated in the same way matches later documents that required both parties to be executed by beheading or hanging (Beirne 1994: 40). Outside England, Bartosiewicz (2012: 223) reports on several deviant burials, whereby humans and dogs were buried in unconsecrated ground. His list includes one eleventh-century burial from Visegrad-Varkert in Hungary where a women's mutilated body was interred with six dogs, an assemblage that has been interpreted as a 'witch burial' but could equally be linked to bestiality, the two frequently being associated.

According to Miletski (2005: 7), the high point for bestiality trials came in the fifteenth and sixteenth centuries. This is coincident with the most significant period of witch hunts, which relied on evidence of bestiality to secure guilty convictions (Beirne 1994: 39). Understandably, the risk of being accused of bestiality or witch-craft – crimes that carried capital punishments – would have made people wary of spending too much time in the company of animals. So, for much of the medieval and post-medieval periods, pet-keeping was not something to shout about. However, like bestiality, it seems almost certain that animals were kept as companions and that the same has been true of human societies for millennia, just with different motivations and social perceptions.

Companion Animals and Pets

Today, pet-keeping is exceptionally widespread within Western society, involving an extensive range of species – from invertebrates to large mammals – and generating huge economic profits for the suppliers of pet products. The pet industry is one of the most important elements of world trade and it is growing steadily as pet-keeping increases in developing countries, as they gain more disposable income and their cultural ideologies change with the shift towards urbanization (De Silva and Turchini 2008). This global movement has severe environmental implications linked to the removal of certain species from their natural range in order to satisfy overseas consumer demand (e.g. see Schlaepfer *et al.* 2005 for impact of over-collection on wild animal populations), the introduction of 'alien species' to new ecological niches (e.g. Calver *et al.* 2011 estimate that a total of 130 million birds and 1.5 billion small mammals are killed annually by cats in the UK and USA), and the supply of foods to maintain the pets in their new habitats (e.g. De Silva and Turchini 2008 have highlighted the impact of pet food production on global fish stock). These important issues are largely overlooked and ignored in the West, a situation that rather characterizes and reflects our egocentric and exploitative culture. In this way, pet-keeping can be seen as a barometer of the relationship between humans and the natural world and, perhaps therefore, it can be examined as a useful proxy for socio-economic and cultural change.

The extreme forms of pet-keeping currently being experienced on a global level are, like our societies, fairly unusual on a broader temporal scale and we should not expect to find similarly overt examples of pet keeping as a specific form of social practice in relation to past societies. To this extent, Serpell and Paul's (1994: 129) definition of a pet as an animal 'kept primarily for social or emotional reasons rather than for economic purposes' is unlikely to be applicable to the majority of cultures examined by archaeologists. However, it is clear that humans and animals are, and have always been, drawn to each other's company (Serpell 1996) and if we can find the evidence for these relationships in the past, we may be able to better understand the mind-set responsible for them. Such a task is not easy. Indeed, research into pet-keeping amongst modern groups, where the humans can actually be interviewed, often yields conflicting interpretations. This is particularly the case with regard to 'pet-keeping' amongst modern hunter-gatherer groups.

Pet-keeping amongst 'tribal societies' is well documented in the historical and anthropological literature; there are many reports of tame animals being kept within hunter-gatherer households where they are raised as named members of the family, and mourned at death (see Erikson 2000; Serpell 1996: 60–72; Simoons and Baldwin 1982). Less consideration has been given to the rationale for these relationships but various explanations have been posited: for instance that they represent live meat stores that can be tapped in times of need, or that they are kept as 'toys' for children so that they might come to know better the animals they must hunt in adulthood.

Erikson (2000) takes issue with these traditional, functional, arguments, instead proposing a more magico-religious explanation. Drawing upon the anthropological literature concerning the morality of hunters and how they must behave appropriately in order to be successful (see Chapter 3), Erikson proposed that the act of killing incurs a worrysome debt to the animal masters. Whilst hunters may employ rituals to appease the animal masters, this may not be sufficient to ward off their retaliation, which could come in the form of famine or disease. Erikson (2000: 14) suggests that, if hunting poisons the relationship between humans and the animal masters, the taming and raising of particular individuals may serve as an antidote. By example, the Piaroa of the Amazon tame and feed parrots in the belief that they sing away diseases sent by the Spirit Masters (Erikson 2000: 12). Such a scenario may also account for the behaviour of the Aino in relation to their capture and maintenance of bears, which they love but also hope will ensure they are protected (see Chapter 6). What sets the Aino apart is that they do eventually eat their 'pet' bears, which contrasts with Erikson's (2000: 21) suggestion that pets are seldom eaten. There may be environmental reasons why some hunter-gatherer groups eat their pets (e.g. those in less stable environments) and others do not (e.g. those where resources are more assured). If this is the case, it would suggest that pet keeping is, perhaps, a fundamental human desire but one that is permissible only in the right 'economic' setting: where people can afford to cherish animals, they will do so, hence the current pet-keeping upsurge in the developing world (De Silva and Turchini 2008).

The possibility that environment and economics may underlie pet-keeping practices has some implications for understanding domestication. As outlined in Chapter 2, regardless of geographical location, the beginnings of domestication tend to coincide with periods of environmental improvement and stability, thus we may also envisage 'economic' stability. Under such improving circumstances hunter-gatherers would not have needed to eat their pets and it is easy to imagine how the keeping of tamed animals could have developed into the maintenance of small domestic herds (Simoons and Baldwin 1982). However, Erikson's (2000: 22) central argument is that, rather than paving the way to domestication, pet-keeping (as a mechanism for appeasing the animal masters) represents a statement of allegiance to the hunting way of life.

In the absence of decent evidence from the archaeological record, these debates are likely to continue in perpetuity, especially since the remains of 'pets', particularly those that were eaten, will be morphologically and contextually indistinguishable from those that were hunted. The only way we could identify pet-keeping in the zooarchaeological record is if we were able to reconstruct human–animal relationships (for it is these that separate the pets from the hunted) directly from the bones. But of course we do have the potential to do this via bone chemistry, with isotope analysis offering an insight into the diet of ancient animals and the ability to differentiate between those that lived with, and were fed by, humans and those that existed outside this relationship. For instance stable isotope analysis of dog (Barton *et al.* 2009) and cat (Hu *et al.* 2014) remains from China have shown how the method can highlight the shifts in human–animal relationships that developed during the earliest phases of domestication.

At present, few such isotope studies focus on human–dog relationships at the level of the individual, but data have been generated unwittingly by researchers hoping to reconstruct human diet. The available data for Britain are shown in Figure 7.2, which presents the average $\delta^{15}N$ results for humans and dogs recovered from the same sites, split by archaeological period. Although the data are very generalized they do highlight that the diet of dogs mirrors that of humans, as suggested by Guiry (2012); however more importantly it demonstrates shifts in the proximity of human–dog diet through time, which is worthy of consideration when viewed in their broader context.

The data for the Mesolithic humans and dogs are interesting as, although this period is considered to be the time that dogs became domesticated, the distance between the human–dog $\delta^{15}N$ averages is quite considerable. It should be noted that the available data for this period are limited but, even so, they may suggest that the human–dog relationship was not that close, at least in dietary terms. The situation appears to change into the Neolithic period, where the $\delta^{15}N$ averages for humans and dogs are much closer, suggesting greater similarity in diet and, possibly, a more intimate, commensal relationship. More data are available for the Iron Age and while the average $\delta^{15}N$ data demonstrate a continuation of the closer human–dog diet, the picture is much more complex when viewed by site type. For

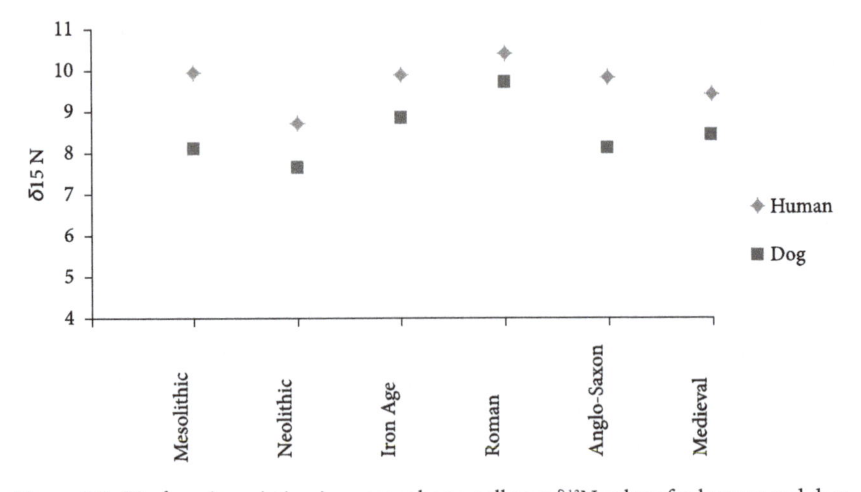

Figure 7.2 Diachronic variation in average bone-collagen $\delta^{13}N$ values for human and dog remains from England.

Sources: Hedges *et al.* (2008); Reynolds (2013); Lightfoot *et al.* (2009); Stevens *et al.* (2010, 2012).

instance, on settlements such as Wetwang Slack in Yorkshire, all dogs have $\delta^{15}N$ values that are substantially lower than, or at the bottom of, the human range (Jay and Richards 2006). By contrast the results from the Danebury Environ project (Stevens *et al.* 2013: 101) indicate greater variety whereby some dogs had markedly lower protein diets than humans but others far more protein rich than many members of the community. This suggests, as was argued in Chapter 1 (page 6), that our classification of animals into categories such as 'dog' does not do justice to the complexities of the human–dog relationships that existed in the past; indeed, it serves to obscure them.

According to Figure 7.2 the transition from the Iron Age to the Roman period was accompanied by a reduction in the distance between the $\delta^{15}N$ values of humans and dogs, suggesting that their diets were closer in this period than at any previous point. The possibility that this reflects increasing intimacy between humans and dogs is consistent with the broader archaeological and historical data. Reviews of documentary and epigraphic evidence for pet-keeping in the Roman world highlight the popularity of dogs as companion animals: they were loved in life and mourned in death, some being given special burials with their own monumental headstones (Bodson 2000; Serpell 1996: 47). No grave stele for pets have been recovered in Roman Britain but Morris's (2008a) study of Associated Bone Groups (ABGs) has highlighted that dog burials become the most common type in the Roman period, accounting for over 50 per cent of all examples (Figure 7.3). It is likely that many of these dog ABGs are not the burials of a cherished pet; for instance Woodward and Woodward (2004) have suggested that

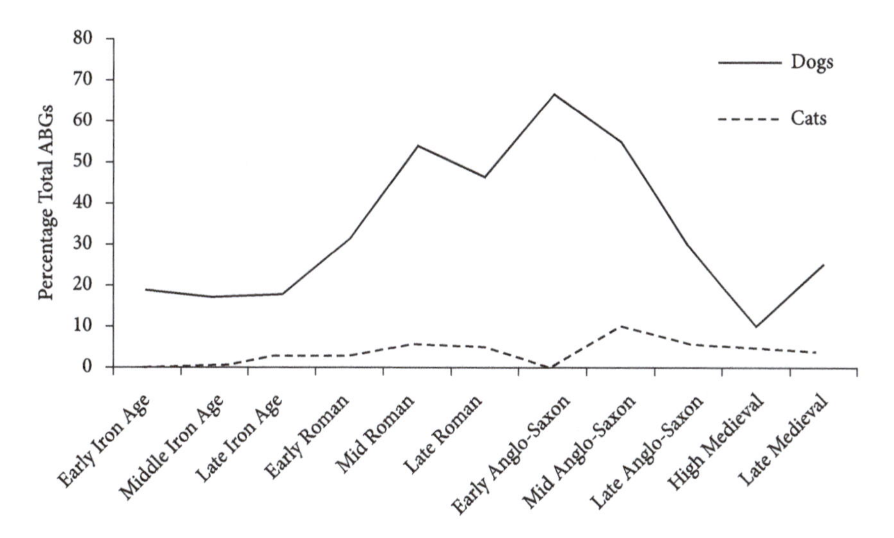

Figure 7.3 Inter-period variation in the representation of dog and cat Associated Bone Groups.

Source: Morris (2008a).

large numbers of dogs recovered from a Roman well in Dorchester reflect sacrificed animals associated with foundation rites. The dramatic rise in dog ABGs does, however, suggest that attitudes to dogs changed from the Iron Age to the Roman period and there are examples where dogs appear to have been given formal burials, such as a small dog that was buried in a human-size grave in fourth-century York (Baxter 2006).

The diversification of dog forms and the appearance of smaller types, like the York individual, have been cited as evidence for the introduction of pet 'lap dogs' (Harcourt 1974; Baxter 2006). However Clark (personal communication) has pointed out that their appearance coincides with the arrival of the black rat (*Rattus rattus*) in Roman Britain (Figure 4.1) and has suggested that rather than representing cherished pets these small dogs may be working ratters. There is, of course, no reason why dogs could not have multi-tasked and fulfilled both roles simultaneously, a point made well by Salmi (2012), albeit in relation to companion animals in post-medieval Norway. In many cultures, animals are not strictly classified as 'pets', 'pest' or 'working animals' and instead, like humans, can have multiple identities that are played out variously in different contexts. Again, in order to reconstruct animal lifeways and understand how dogs were viewed we need to bring together all available data for individual animals.

To date, the most comprehensive life-history reconstruction is that which has been undertaken for a small 'toy breed' dog recovered from the Roman cemetery of Yasmina, Tunisia (MacKinnon and Belanger 2006; MacKinnon 2010c). This particular individual, which was found in the grave of a 10- to 15-year-old human, was about the same age as the human it accompanied (possibly older) and was very sick. It had extensive arthritis and tooth loss that MacKinnon and Belanger suggest would have prohibited its survival in the absence of human care. Isotope data support the idea that the animal was cherished and well fed, its $\delta^{15}N$ values being 12.5, significantly higher than the average for Romano-British dogs (Figure 7.2) and also in excess of average values for Mediterranean dogs, which tend to be less than 11.5 (MacKinnon 2010c: 304). The dog's diet may even have been responsible for the animal's poor health, since high-protein intake has been linked to increased bone loss in animals (Amanzadeh *et al.* 2003). Perhaps this could be an example of where over-feeding of a pet can become a form of affection-borne cruelty.

As in the case of the Yasmina dog, animal palaeopathology can provide good evidence for human treatment of animals. Datasets are currently limited but a few studies on dogs are available and they offer some useful information concerning inter-period variations in attitudes towards animals. MacKinnon's (2010c) survey of Roman dogs from the Mediterranean suggested that less than 3 per cent of animals demonstrated pathological conditions. This low prevalence of pathology contrasts with Teegen's (2005) data for rib fractures exhibited by medieval German dogs, where the average prevalence was 7.7 per cent. Dog-directed violence is also indicated for medieval France: a skeleton recovered from tenth to twelfth-century Guimp, France, exhibited numerous lesions of traumatic origin suggesting that the animal suffered prolonged and/or repeated abuse (Binois *et al.* 2013).

While the palaeopathological data are far from robust, they do point towards a post-Roman shift in attitudes to dogs and this is also indicated by other sources of evidence. For instance the average $\delta^{15}N$ data shown in Figure 7.2 suggests that humans' and dogs' diet became more divorced in the Anglo-Saxon period, and the frequency of dog ABGs declines dramatically over the course of the post-Roman period (Figure 7.3). When all the data are viewed together it seems probable that the trends can be related to the arrival of Christianity and the rise of a worldview that placed humans firmly above animals, stressing separation between the two. In much the same way that bestiality was seen to corrupt the natural order by making humans like animals, pet-keeping was seen to make animals like people: an equal perversion of the Chain of Being.

As with bestiality, pet-keeping was initially tolerated and several early Christian saints were closely associated with pet animals, such as St Guineford who had a pet dog. That early medieval monks were also openly affectionate towards cats is indicated by the ninth-century poem *Pangur Bán*, written by an Irish monk about his white cat. Evidence that cats became increasingly common companion animals in the Anglo-Saxon period is provided by Figure 7.3, which shows a slight rise in the frequency of ABGs, potentially indicating an increase in formal burials. A more convincing example of cat keeping is provided by stable isotope analysis of a cat skeleton from the probable monastic site of Bishopstone in Sussex (England). At this site two cat skeletons were analysed alongside a range of other animals and humans from the site: one of the cats had $\delta^{15}N$ and $\delta^{13}C$ values that suggest it may frequently have been fed fishy treats, whereas the other showed no such dietary signature (Poole 2010c: 16).

It is clear that cats roamed many monasteries and other institutions where books were produced, their presence helping to keep the documents mouse-free. On the one hand these animals were useful partners but on the other, cats can be annoying in their habits and they have the tendency to be very disruptive in working environments. The irritating antics of one animal are preserved beautifully in the fifteenth-century document from Dubrovnik, which shows the inky footprints of one attention-seeking individual (Figure 7.4). Another document, this time from fifteenth-century Deventer in the Netherlands, was urinated on by a cat, prompting the scribe to fill the stain with a curse on the animal and all others that ruin documents (G.B. quarto, 249, fol. 68r Cologne, Historisches Archiv).

Work by Beirne (1994) demonstrates that animals were executed for less serious offences than the ruination of manuscripts, and documentary evidence from across Europe shows that, between the fifteenth and seventeenth centuries, numerous animals were tried and convicted of various crimes. This was, of course, the period of witch hunts and bestiality trials, when pet-keeping was driven underground, particularly amongst the lower classes who were more at risk of accusations of bestiality and witchcraft (see page 139). The elite, however, have always played by their own rules and aristocratic pet-keeping actually began to flourish at around the same time. Later medieval and post-medieval iconography shows an increasing representation of companion animals in

Figure 7.4 Picture the scene as the scribe returned to his work.

Source: Emir Filipovic.

portraiture (Beirne 2013: 190–5; Kalof 2007: 117; Pinault Sørensen 2007) and historical evidence provides many accounts of pet-keeping amongst aristocratic women (Ritvo 1987: 160; Serpell 1996: 49; Walker-Meikle 2012). Dogs were favoured pets but through the course of the medieval period exotic animals – parrots and monkeys – became more common. Zooarchaeological evidence for the keeping of exotic animals has been recovered from a number of high-status sites, including parrot remains from seventeenth-century Castle Mall in Norfolk (Albarella and Thomas 2002: 36) and tortoise bones from post-medieval Stafford Castle, Staffordshire (Thomas 2010).

As animal keeping increased, and the fashion gradually trickled down the social classes, a more positive attitude towards animals developed and historical sources indicate a move away from animal cruelty, with the banning of bear-baiting and cockfighting (Ryder 2000: 55; Serpell 1996: 161; Thomas 1983). Although zooarchaeological data for the early modern period are limited (Thomas 2009), recent analyses of large urban assemblages from nineteenth/twentieth-century Chester (England) have produced sizeable collections of dog remains, particularly from tannery deposits. Analyses of these remains (Clark personal communication) have shown very low rates (0.7 per cent) of trauma-related pathologies, suggesting that the majority of animals were not mistreated. To our mind it may seem paradoxical that animals that were well treated in life might be sent off to the tannery or knacker's yard in death; but at a time when it was widely believed that animals have no immortal soul, the use of their carcasses is entirely logical. In many ways, the industrial utilization of all parts of the animal – their skin, meat (in the case of horses), bones, horn and fat – could be seen as more respectful and environmentally responsible than is the case today, where the concept of animal 'waste' exists.

Summary

This brief review of the evidence for changing attitudes to animal cruelty, bestiality and pet-keeping has revealed how perceptions of these different human–animal relationships have shifted according to prevailing cosmologies. In the Mesolithic period, there is reason to suppose that the hunter-gatherers saw humans and animals as permeable entities and, under such circumstances, sexual relationships between species may have been considered not only permissible but, potentially, essential for the seduction of quarry and the appeasement of Spirit Masters. It may be for the same reasons that animals were initially tamed, with orphaned individuals being raised within families as 'pets'. Such a scenario is difficult to discern from the archaeological record and would require a comprehensive programme of isotope analysis at the level of the individual to try to differentiate the remains of wild animals from those that were fed by people.

The only species that has been the subject of detailed isotope analyses is the dog, and the limited data do not suggest a close dietary connection between Mesolithic humans and their dogs. This situation changed into the Neolithic, when the isotope data suggest that people and dogs became more commensal, as might be expected for a period when humans and animals began to cohabit. The increased proximity between humans and domestic animals must have offered great opportunities for sexual encounters, perhaps explaining some of the human–animal burials found in the Near East, as well as the human–animal hybrid iconography of the period. Hybrid iconography continues into the Bronze and Iron Ages, periods for which bestiality, particularly with horses, has been proposed in relation to kingship rites.

The Roman period marks a moment of change, observable across the datasets considered in this chapter. In terms of iconography, there is a move away from hybrid figures towards the separate depiction of humans and animals. By contrast, the isotope data indicate a greater degree of human–dog commensalism during Roman period and certainly historical and epigraphic evidence point towards the emergence of pet-keeping in this period. Some animals were clearly well treated, as is indicated by the low prevalence of trauma-related pathologies on dog bones, and the care shown to individuals such as the Yasmina dog. On the other hand, while there is a marked increase in the frequency of deliberate dog burials in the Roman period, not all of these animals were cherished pets – some appear to have been sacrificed and deposited for reasons of medicine and magic: in this period the term 'dog' covered a variety of different animals. Combine this with the extreme animal cruelty exhibited within amphitheatres and the Roman period seems paradoxical in its attitudes to animals.

The decline of the Roman Empire was accompanied by a reversion to the pre-Roman worldview. In many geographical areas human–animal hybrids can be seen to re-emerge within artworks, which could be taken as evidence for a greater respect for animals. However, the early medieval period was not without animal-directed violence, and sports such as cockfighting were probably endemic across much of Europe. Anthropological studies remind us that 'animal cruelty' is a cultural stance; many people who engage in blood sports perceive no cruelty in their actions. In some cases the care they lavish upon fighting animals is in excess of that which they give to fellow humans, a scenario that is perhaps suggested by the evidence from the Wien-Csokorgasse cemetery. At this site the relationship between particular humans and particular chickens appears to have been very close, with gender and status-based identities being negotiated and emphasized through human–chicken interactions.

Anthropological studies have shown that cockfighting can border on bestiality and there are other sources of evidence, particularly iconographic and historical, that suggest zoophilia was commonplace in the early medieval period. However, the spread of Christianity eventually saw inter-species relationships outlawed and punished with severe penalties for all parties involved, evidenced by findings from execution cemeteries. Bestiality was believed to corrupt the natural order according to the Christian worldview and, for the same reason, pet-keeping was similarly frowned upon. Evidently, people continued to have companion animals, particularly those who were most immune to being accused of witchcraft and bestiality (ecclesiastics and the secular elite). For the average person, however, the risk of being seen to be too kind to animals may have encouraged the popular culture of animal fighting and baiting that reached its peak in the sixteenth and seventeenth centuries.

By the beginning of the eighteenth century people were beginning to question the morality of inflicting pain upon animals and it was at this time that the likes of Milton, Pope and Isaac Newton commended a vegetarian diet (Ryder 2000: 57). However, their arguments were not widely adopted and meat eating both continued and accelerated to the present day, as will be seen in the next chapter.

Chapter 8

MEAT

Throughout this book I have argued that animals and their remains represent more than just ingredients or sources of protein for human consumption. However, it is an inescapable truth that people eat animals and the majority enjoy doing so. Regular meat consumption, of the high level experienced within modern Western societies, is a fairly recent phenomenon and today's daily purchases of pre-packaged meat of unknown origins would have been unthinkable in the past, or even a few generations ago. For the majority of archaeological cultures that we study, the lives of the people were substantially entwined with the domestic and wild animals with whom they dwelt. And because of the closeness of their relationship, people are unlikely to have been detached from the slaughter of animals (see Chapter 6) or the distribution and consumption of their meat. Anthropological investigations of 'traditional societies' provides a source of arguably more relevant attitudes towards meat, and even a cursory examination of the literature demonstrates that meat is the most highly prized foodstuff whose distribution and consumption are powerful sensory and symbolic acts (Fiddes 1991; Lokuruka 2006; Symons 2002: 442).

Whilst meat is widely acknowledged to be the most esteemed foodstuff, Fessler and Navarrete (2003) have proven that it is also the most proscribed, with no society on the planet eating all the edible animals available to it. Taboos over flesh abound in the anthropological and historical literature (e.g. Fiddes 1991: 132; Simoons 1994) but decisions about which animals are deemed (in)edible are seldom to do with issues of palatability: most things 'taste like chicken' (Quine 1951). Instead they reveal much about a culture's social psychology and ideology. It is not just the choice of 'species' that is important with regard to food preferences and taboos: decisions about how animals should be processed (for instance Kosher or Halal methods) and distributed are equally significant, bound up with cosmology and notions of 'cleanliness', social reproduction, wealth and status.

Over the last few decades, zooarchaeologists have become increasingly adept at identifying social inequality, characterizing 'high-status' and 'low-status' sites based on the presence/absence of different food animals, their age groups, body parts or other variants (e.g. Ashby 2002; Crabtree 1991a). In some cases, human groups belonging to particular religious or ethnic identities have been identified based on the presence/absence of particular species/body parts and associated material culture (e.g. Ijzereef 1989). However, with this advance has come the recognition that social inequality and cultural diversity are more complex than simple static labels. The perception and expression of identities are situational and shifting; what may be a marker of, say, elite or ethnic identity in one setting can be a trait of lower social standing or an entirely different ethnicity in another (deFrance 2009; Jones 1997; Sykes 2004, 2007a; Twiss 2012). This is particularly the case for foodstuffs, as their meaning and significance are usually constructed through the social mechanism surrounding their procurement, distribution and consumption (Hamilakis 2000). This means that the same food, for instance venison, can have contradictory meanings depending on the procurement process (e.g. whether the venison was obtained through legitimate hunting or illicit poaching).

The study of food and foodways is represented by a large and growing literature, and no single chapter can do justice to such a broad field (although see Russell 2012: 358–94, and Twiss 2012, for excellent reviews of current research). Rather than trying to summarize the existing literature, this chapter will use case studies for Iron Age/Roman and, in particular, medieval England to exemplify some of the theoretical and practical approaches that might be adopted in order to elucidate the meaning of meat and its relevance for understanding social dynamics.

Food Taboos and Preferences

Research into food taboos has traditionally separated into two camps: the 'functionalist' and the 'symbolic'; the fault line following the disciplinary boundaries between evolutionary economics and cultural anthropology/geography. The most famous proponent of the functionalist school is Marvin Harris (1974) who argued that taboos are fundamentally logical and can be explained in terms of utility and sustainability, be it environmental, economic or from the perspective of human health. Within this paradigm the Judaeo-Islamic prohibitions against pork was explicated by the arid conditions of the Near East, to which pigs are physiologically and behaviourally maladapted (Harris 1974: 42). Similarly Harris (1974: 21) interpreted the Hindu ban on cattle consumption as reflecting the vital nature of the animals' secondary products for sustaining India's population and agricultural regimes, requiring that they be kept alive in great numbers. These rational arguments have been extensively critiqued (e.g. Alvard 1995; Simoons 1994) but new research continues to suggest that logic and benefit underlie food taboos. Henrich and Henrich (2010) for instance have recently shown that Fijian fish taboos for pregnant and lactating women effectively reduce their chances of food poisoning by up to 60 per cent.

Functionalist theories are certainly compelling but, as Fessler and Navarrete (2003: 10) point out, whilst they may explain social avoidance of a particular foodstuff, they do not account for the development of a cultural taboo – something that requires effort on the part of the community to create, maintain and exact punishment where taboos are broken (see page 152). This is where the symbolic paradigm seems more appropriate, situating taboos within a network of ideological associations that are attached to particular animals. This school of thought, pushed forward by the likes of Claude Lévi-Strauss (1969) and Mary Douglas (1966), saw food taboos developing where animals were considered either 'sacred' (e.g. totemic animals that should be above consumption) or 'profane' (animals that were considered inherently 'dirty' or symbolically contaminated by association with an 'other' cultural group and so whose meat would be polluting for the consumer). The associations themselves are seen as being culturally specific, dependent upon the classification systems imposed by people (Joy 2009). In this way Douglas (1966) argued that the Hebrew proscription of pork is related to the pig's deviancy from the categorical system outlined in *Leviticus* 11.4 – that pigs have cloven hooves like other ungulates but do not chew the cud, so are ambiguous and therefore dangerous. An abhorrence of boundary crossing is certainly indicated by most monotheistic religions, as reflected by their attitudes to bestiality, pet-keeping and hybrid gods (Chapter 7), and would also account for the Biblical injunction against the consumption of water birds and amphibians – animals that transgress the elements of earth, water and air.

Issues of classification could explain why animals that are seen as too close to humans (e.g. pets) frequently become the subject of taboos (Fiddes 1991: 132–43; Joy 2009); however, there are alternative views. Evolutionary and biological scholars favour the idea that it is linked to the spread of pathogens, which are more likely to transfer between closely associated species. By contrast the cross-cultural work of psychologists such as Ruby and Heine (2012: 47) emphasizes that humans are disgusted at the thought of eating an animal that they consider possesses mental and emotional capabilities akin to their own. Certainly amongst the Suri, where individual humans are linked to individual cattle (with both individuals sharing the same name), a human would never eat the meat from its cattle name-sake, nor that from an animal of similar name or coat colour: that is considered 'cannibalism' (Abbink 2003: 348).

Fessler and Navarrete (2003) have conducted the most comprehensive examination of food taboos to date, examining the pros and cons of the different paradigms. Whilst they accept the importance of an animal's symbolism in sculpting responses to the idea of its consumption they suggest that, as with purely functional explanations, symbolism alone cannot account for all meat taboos. Instead they advocate a more holistic psycho-social approach that takes account of all of the differing theories. They propose the emotion of disgust to be the spark responsible for the rise, spread and persistence of taboos. Their paper is essential reading for anyone interested in food taboos, but it can be summarized as follows. Fundamental to Fessler and Navarrete's (2003) hypothesis is the principle that, as omnivores, humans are particularly drawn towards meat but, because of this, they are at risk from ingesting a wide variety of toxins. As such, humans have evolved a predisposition to being both attracted to, but also wary of, meat and feeling disgust towards it, especially meat derived from new and unknown sources (neophobia). Fessler and Navarrete (2003: 16) propose the evolution of 'socially mediated ingestive conditioning' to have been an important process in the development and transmission of life-preserving food habits. This behavioural trait, which is probably responsible for the success of the human species, allows whole communities to rapidly acquire food aversions vicariously by, for instance, witnessing an individual vomit after ingesting a meat that induces nausea. This response occurs due to a second process that Fessler and Navarrete (2003: 15) term 'egocentric empathy', another evolved trait whereby humans feel personal fear and disgust by witnessing it in others. Together, these two processes can create a situation of 'normative moralization', where a community believes that their behavioural norms are correct and that anything outside this is somehow wrong or disgusting: to explain this Fessler and Navarrete (2013: 14) use the example of handedness, and the fact that people deviating from the right-handed norm are viewed as sinister.

Beyond these deep-seated evolutionary processes, there are many more superficial psycho-social mechanisms that have the potential to convert meat avoidance into an institutionalized taboo. For instance socially mediated ingestive conditioning and normative moralization lend themselves to manipulation and exploitation by the social elite due to the universal human emotions of shame and pride (Fessler and Navarrete 2013: 17). These emotions give people a heightened awareness of, and a desire to conform to, wider patterns of behaviour: as Fiddes (1991: 34) has pointed out, 'we feed not only our appetite but also our desire to belong'. For this reason, humans will readily imitate the beliefs and practices of others that they admire and, as a result, it is possible for individuals of high standing to erect taboos that will serve to enhance their own resources or power (Fessler and Navarrete 2013: 17). This is particularly the case where taboos can be linked to group identities or religion, itself a mechanism for power generation and social control (Graham and Haidt 2010; McCullough and Willoughby 2009; Saroglou 2011).

Whatever the reasons behind them, food taboos and preferences are fascinating and because they have the potential to leave discernible traces in the archaeological record, they deserve attention from zooarchaeologists. Considerable research has been undertaken with regard to pork prohibitions (Hesse 1990; Hesse and Wapnish 1997; Ijzereef 1989; Lobban 1994; Redding 1991) but beyond this topic, most studies of taboos have ignored the archaeological data. This point is made well with regard to the taboo over horseflesh in Christian medieval Europe, for while many scholars have examined the historical and iconographic evidence (e.g. Simoons 1994) few have considered the horse remains themselves.

Recently the situation has been redressed by Poole (2013) who provides a comprehensive survey of the zooarchaeological evidence for the representation and butchery of horse bones in Anglo-Saxon England. He demonstrates that rates of butchered horse bones were fairly high, at 27 per cent, in the Pagan Early Anglo-Saxon period (fifth to mid-seventh centuries) but dropped substantially to 13 per cent with the arrival of Christianity to England in the Middle Anglo-Saxon period (mid-seventh to mid-ninth centuries). When the Early Anglo-Saxon data are viewed together with the contextual evidence – in particular finds of possible feasting deposits that contain horse remains – they would appear to suggest that low-level horse consumption did occur in Early Anglo-Saxon England. Given that horses were highly prized, probably cultic animals at this point (Fern 2010), it seems possible that the eating of horse flesh was undertaken on ceremonial occasions in the belief that the animal's properties would be transferred to the consumer (Fiddes 1991: 67, 177). Such practices are thought to have been widespread in the Iron Age as a form of power-conferring kingship rite (see Chapter 4, page 83). If the same were true for the Anglo-Saxon period, we might expect that horse consumption would become conspicuously abhorrent to a new group that was attempting to establish its own mechanisms for power generation and transferral. Certainly historical evidence, in the form of papal letters and penitentials, indicates strong Christian opposition to horse-meat consumption, the practice being vilified as an expression of paganism (Simoons 1994: 187). As Poole (2013: 320) points out, the introduction of Christianity offered new opportunities for the elite to legitimize their power through display. As an 'exotic' religion that was initially associated with the elite, Christianity would have carried enough social cachet to deploy socially mediated ingestive conditioning and normative moralization resulting in a taboo over horse meat consumption. The social trickle-down of the taboo is demonstrated well by Poole (2013: 329) whose data suggest that horse consumption endured longer amongst the low-status rural population, consistent with the idea that these social groups tend to be more conservative and resistant to religious and cultural change (e.g. Cool 2004: 28–30; Insoll 2004: 100).

In theory it should be possible to detect the shift in attitudes to horse consumption via stable isotope and lipid analysis of food residues in pottery sherds. Work by Outram *et al.* (2009) was able to identify and differentiate between horse adipose and dairy fats preserved in Bronze Age ceramics from Kazakhstan and the same methods have been applied to pottery from a wide range of English sites dating from the Neolithic to the medieval period. Evershed *et al.*'s (2002) review of these studies concentrates on the evidence for ungulate milking in Prehistory; however, the data presented in the article are shown against reference data for horse fats and none of the archaeological samples – including those from Middle Anglo-Saxon Wicken Bonhunt (Essex) and Late Anglo-Saxon West Cotton (Northamptonshire) – yielded results consistent with the cooking of horse products. This negative evidence would seem to support the absence of horse consumption after the arrival of Christianity but more analyses of the Pagan period (Early Anglo-Saxon) ceramics would be required to bring the results into relief.

Whilst neither the Wicken Bonhunt nor West Cotton assemblages yielded evidence for horse consumption, ceramic samples from both sites indicated the presence of poultry fats, consistent with the consumption of chickens or geese. This is interesting considering the complete lack of evidence for poultry fats in the Iron Age and Roman ceramic samples considered by Evershed *et al.* (2002): none of the data from Iron Age Yarnton (Oxfordshire) or Iron Age/Romano-British Stanwick (Northamptonshire) indicated the consumption of poultry. These results are, perhaps, unsurprising given the zooarchaeological and historical evidence both suggest that chickens and geese were not eaten during the Iron Age, possibly due to a cultural taboo (Chapter 4, page 86). If chickens were avoided in the Iron Age, it seems possible that this was due to both neophobia, with chickens being an entirely new form of animal, but also because their exotic status likely rendered them sacred and therefore above consumption (Chapter 4, page 84). By contrast to the horsemeat taboo that arose under Christianity, the arrival of Roman influences appears to have brought an end to the Iron Age taboo against chicken meat. Maltby's (1997) paper on the zooarchaeological evidence for the spread of chicken consumption indicates that the new fashion emerged in Roman towns and forts (settlements with the greatest concentration of 'Roman' occupation) and took longer to spread to low-status rural sites, another example of how socially mediated ingestive conditioning can come into conflict with the normative moralization of low-status rural communities.

Whilst both the rise of the horsemeat taboo and the decline of that associated with chickens can be seen as reflecting the dynamics of cultural change, it is important to recognize that meat taboos and preferences can operate at a number of different social and temporal scales. They need not be permanent or applicable to all individuals within a community. As with the Fijian fish taboo for pregnant and lactating women (Henrich and Henrich 2010), taboos have the capacity to become markers of identities that change through the life-course. This is particularly the case in societies that subscribe to elemental theory, where it is believed that individual temperaments change through life and must be kept in balance by consuming foodstuffs of appropriate humour (see Chapter 1, page 18). For instance, in the medieval period not all animals were deemed acceptable foods for all members of society, with some animals being prescribed or proscribed for particular individuals depending on the perceived temperaments of all parties.

This situation is exemplified well by the wild boar, the violent *bête noir*. On the continent their remains are often found in high frequencies on settlements belonging to the secular aristocracy, who hunted and consumed these fierce animals to embolden themselves and increase their military prowess (Laurioux 1988; Yvinec 1993). By contrast, wild boar are seldom found in assemblages from religious houses, presumably because ecclesiastics avoided animals whose capture and consumption might induce an aggressive temperament. It may be for similar reasons of transference that ecclesiastics appear to have avoided the consumption of partridges, which are scarce on monastic sites until the later medieval period, despite being abundant on secular sites of high status (Sykes 2004). According to medieval bestiaries partridges were thoroughly untrustworthy and so wanton that they 'ignored the law of nature and subjugated beaten rivals to homosexual acts' (Payne 1990: 76). With such a character it is perhaps understandable that their consumption was not actively encouraged within monastic communities (Sykes 2004).

Instead of partridges and boar, assemblages from religious houses consistently yield the remains of roe deer, hare and crane in higher frequencies than is seen on any other site type – these animals were deemed to be meek in character, the roe deer in particular viewed as chaste and pious, and therefore fitting foods for men of the cloth (Sykes 2007b; Yvinec 1993). Fish were also considered the ideal fare for any good Christian, since their flesh provided the ultimate in cold/moist humours, perfect for quenching any trace of heat that could lead to carnal desires (Scully 1995: 59). It is, no doubt, for this reason that monks consumed fish in considerable quantities, as is evidenced by both the zooarchaeological and historical record (Serjeantson and Woolgar 2006). The same is indicated by isotope analysis of human remains, work by Müldner and Richards (2005, 2006) suggesting that fish were a prime source of protein in the diet of later medieval monastic communities.

The influence of Christian food beliefs on secular diet can sometimes be highlighted using stable isotope analysis, as in the case of Barrett and Richards's (2004) study of human remains from early historic Orkney, Scotland. They found that samples from both male and female 'Christian' burials indicated significant fish consumption, whereas those from the earlier (Viking Age) pagan burials suggested fish-eating to be a predominantly male activity: Barrett and Richards proposed that, for the Vikings, fishing and fish consumption were important constituents of Scandinavian masculine identity. It is this ability to reconstruct broad patterns of diet at the level of the individual that renders isotope studies such a powerful tool, as they have the potential to highlight gender- or age-based differences in consumption practices.

Nehlich *et al.*'s (2011) multi-isotope study of human remains from Roman Oxfordshire is a good example of this. By examining $\delta^{13}C$ and $\delta^{15}N$ in conjunction with $\delta^{34}S$ (sulphur isotope ratios, that allow the contribution of marine/freshwater/terrestrial food sources to be estimated) they were able to propose that children aged between two and four years were fed significant quantities of freshwater fish, perhaps as a weaning food, before they moved onto more terrestrial-based foods. Nehlich *et al.* (2011: 4973) state that is it 'difficult to speculate' about the reason for this infant diet; however, perhaps humoural principles offer some explanations. According to the Roman physician Galen (*c.* AD 129–200) children have a moist temperament (see Figure 1.6), and perhaps it was deemed that they should be fed foodstuffs with similar properties (e.g. fish, which Galen specified to be 'phlegmatic') in order that their humoural balance might be preserved (Grant 2000: 15).

In reality, it should not be that surprising that people in the past ate different things at different stages of their life. However, it is not simply the overarching kinds of food that can change according to life-stage but also the type of portions, the provision of which is frequently subject to social and cultural rules.

The Rhetoric of Portions

As set out in Chapter 6, we should expect that, for most of the archaeological societies that we study, the killing of an animal for food would have been an important event. Furthermore, in periods and places where the concept of meat retail was absent, the easiest way to utilize the considerable amounts of meat produced by a single carcass would have been to share it through feasting. However, the breaking and sharing of an animal carcass is undertaken not simply for logistical reasons; participation in communal consumption acts as a strong statement of shared ideology and group identity (Fiddes 1991: 38; Lokuruka 2006). Also, as animal carcasses are by nature hierarchical, the distribution of meat often plays an important socio-political role, with cuts of different (perceived) 'quality' being given to individuals as a meaty symbol of their social position, age, gender, wealth or power (Lokuruka 2006; Schuman 1981; Sykes 2010b; Symons 2002). Exclusion from such performances can be equally expressive, communicating separation and social difference, which may in turn lead to the creation of alternate subversive mechanisms of meat redistribution (Fiddes 1991: 39; Sykes 2007a).

Although the subtleties of meat redistribution practices are not cross-culturally uniform, meat-sharing is an overarching form of social intercourse. It may be used in negotiations between living individuals, which are played out whenever people share a meal together but particularly during feasting situations: there is a huge literature on the (zoo)archaeology of feasting, particularly in relation to Mediterranean Prehistory (Halstead and Barrett 2004; Hitchcock *et al.* 2008; Wright 2004) and readers should refer to Dietler and Hayden's (2001) volume for a range of archaeological and anthropological perspectives on feasting. Alternatively negotiations may be between the living and the dead. For example mortuary feasting has been a feature common to societies from the Bronze Age Aegean (Hamilakis 1998) to medieval England (Lee 2007), and cuts of meat are frequently found in funerary deposits throughout time and space, suggesting provision for the deceased (Green 1992: 105–8; Kroll 2013; Sánchez Romero *et al.* 2008; Williams 2007). Supernatural beings, or their earthly representatives (e.g. priests), may also be placated through meat provision, which may account for the burnt food remains recovered from many sanctuary deposits (e.g. Green 1992: 108–10; Hamilakis and Konsolaki 2004). However, it is important to recognize that all of these different scenarios are permeable – their social, cultural and ideological meaning seeping from one to the other producing considerable overlap, which makes it near impossible to categorize motivations for meat-sharing. Dietler and Hayden (2001) attempted to do precisely this, classify feasts by socio-political motivation. Initially, their classification system was very influential and has been frequently replicated in the literature but, more recently, the idea that commensalism can be neatly boxed by meaning has lost ground as the complexities of meat-sharing become more apparent: see for instance Tsoukala's (2009) inter-disciplinary study of meat redistribution in ancient Greece, which demonstrates how involved the ideology surrounding food sharing can be.

Inter-disciplinary approaches provide perhaps the best opportunity for understanding the mechanics and motivations of meat-sharing and, in an attempt to demonstrate this, I shall rehearse the work that I have published elsewhere (Sykes 2005, 2007a, 2010b, in press) regarding venison distribution in medieval England.

The Dynamics of Venison in Medieval England (Mid-seventh to Mid-fourteenth Centuries AD)

In Middle Anglo-Saxon England (mid-seventh to mid-ninth centuries), much of the economy was based on the accumulation and redistribution of food. Historical evidence for the mechanics of the system is limited but, in general, it would seem that landholders were paid in kind for the use of their land, with portions of these food rents being given over to support the peripatetic kings and their court as they toured their kingdoms. Kings could, in turn, transfer accrued provisions to religious institutions that, unlike the itinerant royal court, were stationary and depended on supplies gravitating towards them. Lower down the social scale, estate workers could expect to receive food payments in return for their services (Faith 1997: 1–4; Lee 2007). A set-up of this kind is difficult to examine zooarchaeologically but it would certainly account for the inter-site variations that have been observed in deer body-part patterns for this period (Sykes 2010b). Figure 8.1 shows that deer assemblages from religious houses show a dominance of meat-bearing elements from the fore-limb but also large numbers of foot bones, suggesting that ecclesiastics were taking receipt of pre-butchered joints of venison and possibly skins (indicated by the high representation of feet). It is feasible that these were gifted by the king or local nobles in return for pastoral care. The over-representation of heads on elite sites finds resonance with the practices of modern hunting and pastoral societies, where crania are frequently conferred with special significance: amongst the Ngarigo of Australia and Turkana of Kenya, for instance, crania are seen as representing the animal in its entirety, and are either claimed by the head of the community or returned to the individual who 'donated' the animal for consumption (see Table 1.1; Lokuruka 2006; Symons 2002: 442).

If heads were deemed to represent high status in the Middle Anglo-Saxon period, the lower limb bones recovered from rural sites may reflect the lower social position of these settlements and their occupants. Nevertheless, it is important to recognize that the inhabitants of these rural sites actually had a position *within* a community, something that was of fundamental importance in Anglo-Saxon society: Old English literature is preoccupied with the concept of community and frequently uses the imagery of the feast hall to express ideas about the maintenance of social order and rule (Magennis 2006). The coming together in a hall to collectively consume the body of a single deer would have been an important occasion, binding the participants together whilst simultaneously defining their social position through the allotment of specific portions. It may be for this reason that in the story of *Beowulf*, the king Hrothgar names his great feast hall '*Heorot*', the hart.

Marvin (2006) has highlighted the significance of this name, arguing that *Heorot* would have carried real meaning, demarcating the hall as a masculine space and symbolizing it as an arena for the cutting up and sharing of venison – rituals that would have been the food-based equivalent of the gift-giving that took place within the hall, where men pledged service to their lord or king in return for weapons, treasure and land (Härke 2000: 397). Historical evidence indicates that

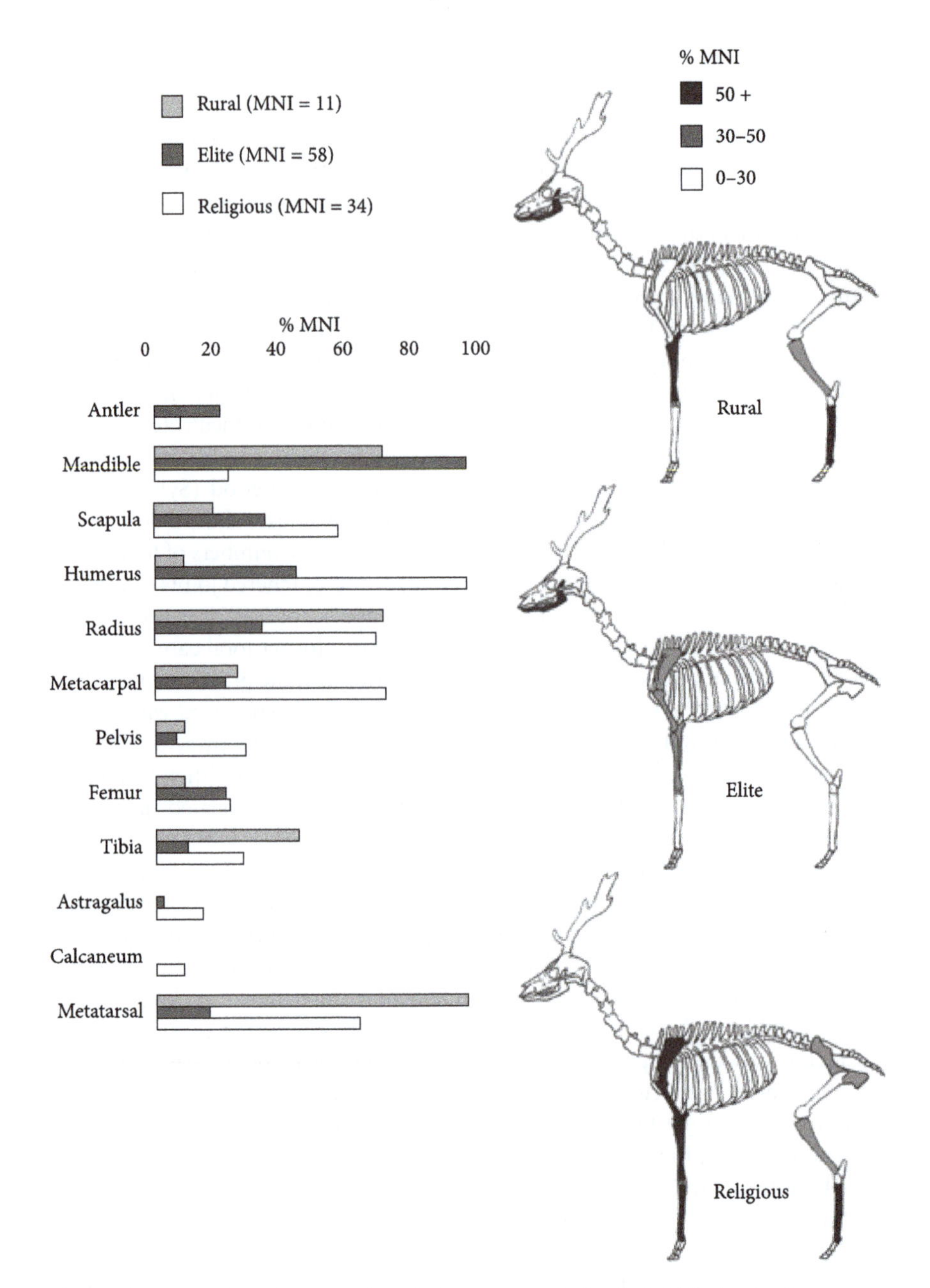

Figure 8.1 Inter-site variation in deer skeletal representation in Middle Anglo-Saxon period assemblages.

Source: Sykes.

generosity in gift-giving was deemed to be the mark of a good leader and, although Anglo-Saxon texts are entirely silent on the issue of food (Magennis 1999), it seems likely that open-handedness was equally desirable in terms of meat provision. Indeed, the importance of the leader as a supplier of sustenance is indicated by the etymology of the word 'lord', which has been traced to *hlafweard*, meaning 'loaf-keeper' (Schuman 1981: 71).

In a situation where the control and redistribution of foodstuffs were equated with power and authority, it stands to reason that the knives physically responsible for cutting up and sharing may have become iconic in their own right, a symbol that their owners possess both resources and the power/generosity to divide and redistribute them. It may be no coincidence, therefore, that at the point we see the appearance of deer body-part patterns indicative of venison redistribution, we also see the emergence of a fashion amongst elite men for particularly long knives, which were worn in a prominent position, located on the belt where other display items were suspended (Härke 1989; Owen-Crocker 1986: 43–8, 100–1).

In Chapter 5 (page 100) I argued that human–animal–landscape relationships are inter-dependent and, in Middle Anglo-Saxon England, just as venison was carved up and redistributed as a currency of social obligation, so too was the land. Indeed, I would suggest that the whole culture of Middle Anglo-Saxon England was structured and defined through the cutting up and redistribution of material wealth. Through the course of the period, territories were increasingly ceded to the Church and secular lords, who in turn leased smaller parcels of land in return for service (Faith 1997; Hooke 1998: 54). Unlike meat, however, the giving of land is unsustainable. Although attempts were made to 'recycle' land, by granting it for a limited period (usually between one and three lifetimes only), large estates were gradually broken into smaller and smaller pieces (Godden 1990: 41). As land and power were carried permanently into the hands of a newly emerging aristocratic class, this set in motion the transformation of Anglo-Saxon society, evidence for which is clear in the zooarchaeological record.

As was seen in Chapter 3 (page 71) the new elite went to great lengths to maintain their social position through ostentatious displays of hunting and, by the Late Saxon period (mid-ninth to mid-eleventh century), hunting had become less a performance of group identity and more a display of royal or thegnly power: a demonstration of ability to muster resources and manpower. Undoubtedly the lower classes were still involved with hunting but their role within the hunt seems to have moved from centre to periphery and they were now obliged legally to act as game beaters for their lord (Faith 1997: 102–3, 111–12). Skeletal representation data for Late Anglo-Saxon deer assemblages supports the idea that the lower social echelons were increasingly excluded from elite culture (Figure 8.2). By contrast to the preceding period, there is little evidence that venison was redistributed through society, and instead all parts of the deer skeleton become equally represented, often in high frequencies, only on elite sites. The privatization of wild resources is mirrored by the privatization of the hall, with Gautier's (2006) study of Anglo-Saxon feasting concluding that hall-based activities became less accessible and more hierarchical during the Late Anglo-Saxon period. These changes were

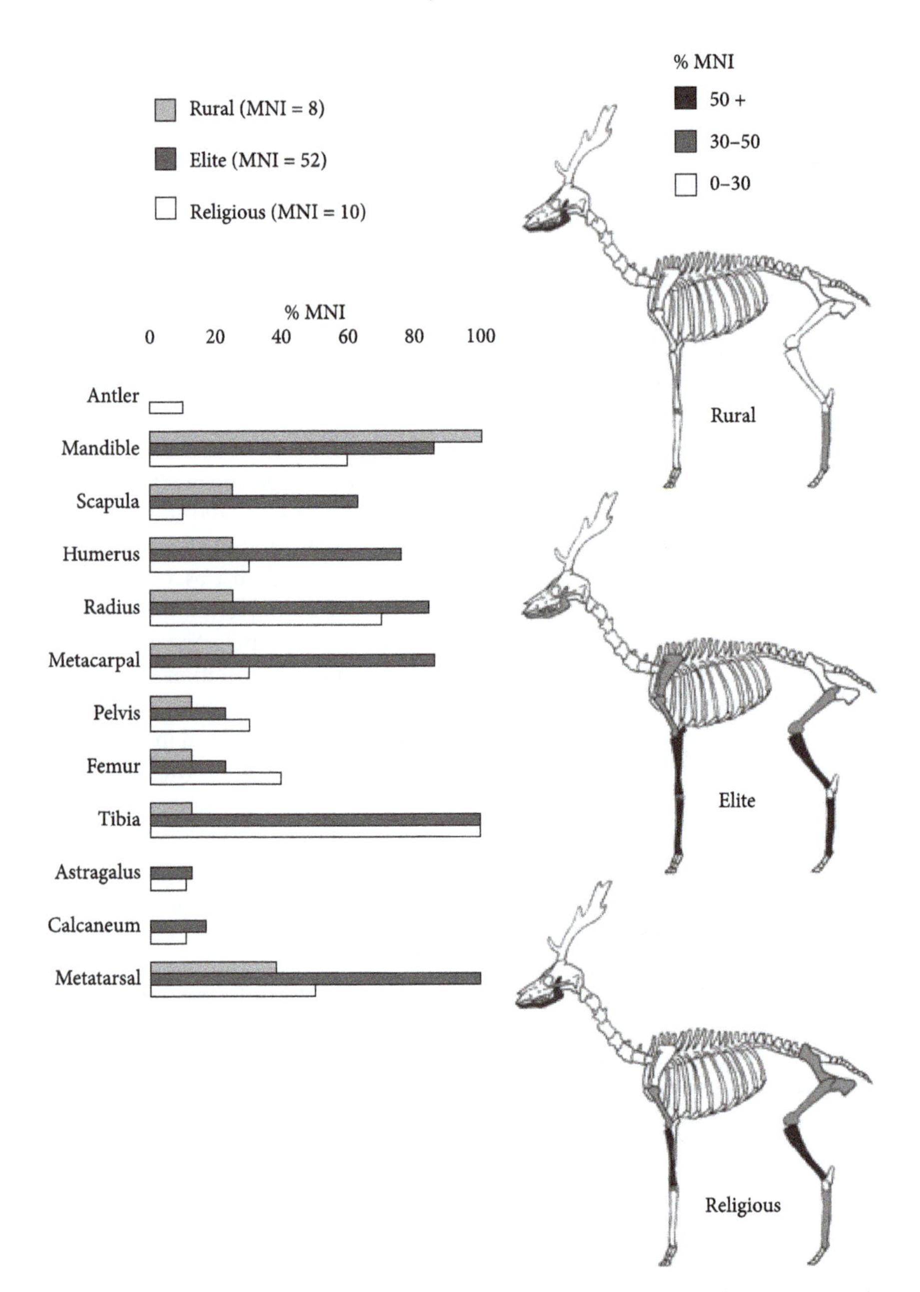

Figure 8.2 Inter-site variation in deer skeletal representation in Late Anglo-Saxon period assemblages.

Source: Sykes.

seemingly concomitant with wider transformations in the system of food rents – by the tenth and eleventh centuries food rents were increasingly being commuted for cash payments, so releasing landlords from the responsibility of hospitality and the public feasting it entailed (Stafford 1980). As the market economy developed, the functional necessity of communal feasting (as a mechanism for meat redistribution) would have been reduced as aristocratic households could obtain meat as required from urban markets. More importantly, the need for feasting as a performance of social obligation and community order would also have diminished because, by the Late Anglo-Saxon period, everyone knew well their place and their duties – these were laid out formally within charters and documents.

It is interesting to note that in the period when food rents and meat redistribution were being abandoned in favour of monetary exchange, there is less tangible evidence for the symbolic significance of the knife. Certainly depictions of knives are rare in Late Saxon art; for instance, although dress is illustrated in great detail in the Bayeux Tapestry, (the 70 m long textile that depicts the events surrounding the Norman Conquest of England), knives are not shown as being part of the attire: they do not hang from belts where they might be expected. In the tapestry the only clear depictions of knives are in a dining scene where three are shown laid at table, perhaps suggesting knives were beginning to be viewed as cutlery rather than personal appendages (Owen-Crocker *et al.* 2004: 251). This would seem to suggest that the 'loaf-keeper's' ability to provide and redistribute resources in general and meat in particular was less of an issue now that power was displayed through social exclusion.

By the Norman period (mid-eleventh to mid-twelfth century) it was an established rule that deer were royal property and that venison was priceless – a perk of office or something that was gifted as a demonstration of royal or aristocratic largesse – it certainly should not be bought or sold (Birrell 1992: 114). As was seen in Chapter 3 (pages 71–73), the historical evidence for the gifting of venison correlates exceptionally well with the zooarchaeological data, with particular portions and sides of venison being given to particular people (Sykes 2007b). Forest workers and parkers in particular received large quantities of venison as a requisite of office (some assemblages from parkers' residences contain upwards of 30 per cent deer bones) and it must be questioned to what extent the parkers would have viewed venison as a 'special treat' when it comprised their daily fare. At the lord's table, however, venison retained its cachet and the meat was redistributed according to the status of the consumer. The prized liver and testicles were reserved for the lord but persons of lower standing were offered offal, or 'umbles'; the saying 'to eat humble [umble] pie' is derived from the social humiliation attached to the consumption of these poorer cuts (Goody 1982: 142). Once again, it is important to recognize that inequality should not always be equated with social division because although it is clear that venison was being used to define social position, its communal consumption must simultaneously have served to create community.

Despite this, there were sections of society who were excluded from hunting or objected to the social group that practised it. For those not permitted to hunt, poaching provided an exciting alternative, with the distribution and consumption of ill-gotten venison carrying its own cachet. By subverting established systems of venison distribution it was possible for the disenfranchised to empower themselves by undermining feudal control. Historical evidence demonstrates that underpaid and disgruntled forest workers were occasionally caught selling their share (and more) on the urban black market (Birrell 1982: 16; Manning 1993: 28–32). This is supported by the archaeological record, with urban deer bone assemblages exhibiting random body-part patterns that bear no resemblance to the ordered skeletal representation indicative of legitimate hunting. The patterns for urban sites must surely reflect the actions of the poaching gangs that operated out of urban taverns, where they also consumed and sold their bag. Within the safety of the alehouse, the communal consumption of their ill-gotten venison, together with the drinking, storytelling and general bravado it entailed, would have cemented the fraternity in much the same way it did for legitimate hunting groups.

This rapid review of the evidence for medieval venison redistribution and consumption has sought to demonstrate that the meaning attached to this meat was dynamic, varied, multifaceted and context-dependent. Indeed, traditions of meat redistribution actually allowed venison from a single carcass to be consumed within a variety of social scenes – some at the aristocratic table, some in lower-status settings and, if gifted portions were sold on the urban black market, it was even possible for venison to move from the legitimate to the illicit. This is the level of complexity that we should anticipate, and look for, in the archaeological record. My work has focused on non-commercial redistribution of a socially prestigious wild meat that was obtained through social and politically charged hunting events; however, this is not to suggest that commercial redistribution of farmed meats is devoid of meaning. Indeed, there is much potential examining the social impact and significance of market-based meat redistribution.

Butchery and Carving

For the UK, Grant (1987) and O'Connor (1982) were the first to show how evidence from butchery marks could be used to pinpoint the rise of professional butchers. By studying multi-period assemblages, Grant was able to chart how practices changed through time: Anglo-Saxon period assemblages demonstrated seemingly haphazard disarticulation by knife, whereas medieval assemblages exhibited more consistent cleaver-based methods of carcass reduction. Included within this shift was a move towards the splitting of carcasses into equal sides, evidenced by the appearance of sagittally cleaved vertebrae. Such butchery practices require suspension of the carcass, and it has been argued that, in the case of cattle, this could have been achieved only with the use of specialist equipment and purpose-built premises — the presence of cleaved vertebrae therefore indicates the presence of professional butchers.

Taking a more integrated approach to the study of medieval butchery, Seetah (2007: 27) demonstrated that the establishment of the Butchers Guild (founded officially in the 1300s but almost certainly existing in some form prior to this) instigated a refinement in butchery style, facilitated by the production of new, specifically created butchery tools. This emergence of artisan butchers would have enhanced meat as a symbol of social differentiation because there were now 'correct' methods, learnt by butchers over a 7-year apprenticeship, for the preparation of different animals. Subscription to these ideals and the consumption of meat prepared in the accepted fashion therefore demonstrated a person's social standing.

Zooarchaeological evidence for butchery etiquette has been identified at a number of Norman period sites, particularly those of high status. The eleventh/twelfth-century assemblage from Carisbrooke Castle on the Isle of Wight, for instance, contained a peacock foot bone exhibiting cut marks produced when the bird's toes were delicately removed from the tarsometatarsus (Serjeantson 2006: 143). Another example, this time of a crane foot bone, was recovered from Norman Lincoln (Figure 8.3). Practically, it would have made more 'sense' simply to remove

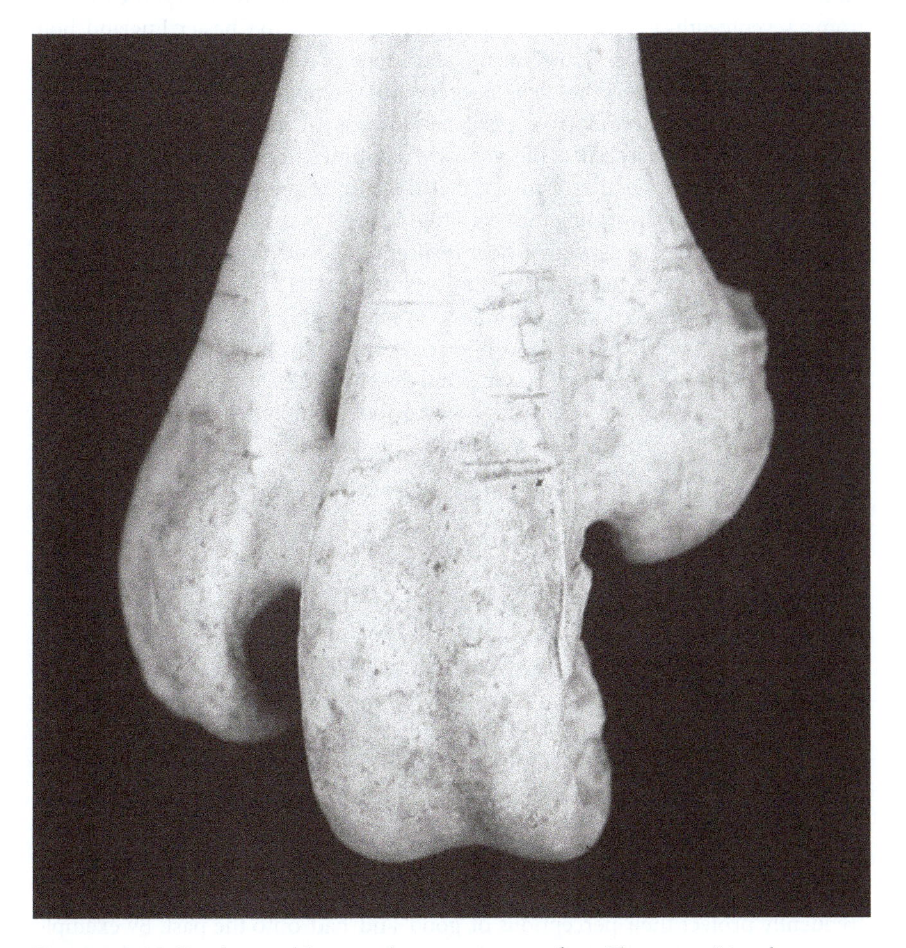

Figure 8.3 Medieval crane (Grus grus) tarsometatarsus from Flaxengate, Lincoln, showing fine cut marks on its distal end – evidence of dressing?

Source: Terry O'Connor.

the entire lower leg, which carries no meat, rather than individually removing each toe, and it must be assumed that the butchery process was influenced by factors other than efficiency (O'Connor 2007: 7). Indeed, it seems probable that this unnecessarily complex method of carcass preparation was linked to the methods of 'dressing' and carving game birds that are documented in later medieval recipe books such as Wynken de Worde's (1508) *The Boke of Keruynge* as matters of patriarchal display.

The social significance of carving, which became particularly fashionable between the sixteenth and nineteenth centuries, is an entirely under-researched area by zooarchaeologists. This is perhaps because most post-medieval assemblages are discarded by archaeologists as 'too modern' (Thomas 2009). Where domestic assemblages are available, even single bones can provide charming insights into family dining practices. For instance, I recently examined a sheep scapula and goose sternum from a seventeenth- to nineteenth-century urban dwelling in Chester, England; both exhibited repeated linear knife marks consistent in direction with the carving instructions described in the contemporary *Mrs Beeton's Household Management* (1861: 360, 504). Mrs Beeton wrote specifically for women, suggesting that the Chester examples of carving may reflect the actions of a dutiful wife and/or mother, although if this were the case we may envisage that the lady of the house plated up in the kitchen rather than performing the task at the table (Symons 2002: 446).

Shifting attitudes to carving – how it should be done, where and by whom – have been considered in detail by the likes of Elias (1978), Goody (1982) and Mennell (1985). These detailed historical studies reveal how etiquettes of meat redistribution changed and evolved, often over very short periods of time – weeks, months and years. This level of temporal resolution is not often available for zooarchaeologists but we need to be aware that fashions change, and it is the motivations for the changes that we should be interested in.

Matters of Taste

Taste is largely a cultural construct. Physiologically the human mouth is able to detect just five tastes – sweet, sour, bitter, salty and umami – so most of our beliefs about what tastes good or bad have little to do with oral senses and more to do with ideology (Mennell 1985). Essentially, it is our worldview that shapes our perception of a food's edibility and desirability (Fiddes 1991: 32). As we have seen throughout this book, human worldviews are in a constant state of dynamism, ever reshaping and reforming in response to individual, social, political and cosmological shifts in perspective. When dealing with matters of taste, the only thing we can be certain of is that tastes change. Whilst this may seem a statement of the utmost banality, zooarchaeologists are surprisingly good at overlooking the truism, and frequently project their perceptions of 'good' and 'bad' onto the past. By example, the widespread belief that insects are 'not food' must be responsible for the dearth of archaeological studies of insect-eating, despite the ubiquity of this practice in non-Western cultures (Sutton 1995; Van Itterbeeck and van Huis 2012).

On the occasions when zooarchaeologists are able to link a particular foodstuff to a particular social group (e.g. the consumption of sturgeon by the Anglo-Saxon elite or the ancient Egyptian preference for pork) we congratulate ourselves on our ability to characterize assemblages using 'vertebrate signatures' (Dobney and Jaques 2002; Sykes 2007b). We then stick fast to our characterizing criteria (e.g. sturgeon = elite) in a way that is more conservative than the societies we study. Very often zooarchaeologists apply animal identity-markers across vast periods of time, such as the 1,000 years represented by the medieval period (e.g. Ashby 2002; Albarella and Thomas 2002; Sykes 2007b). This is despite the fact that, as was seen in the previous section, the social, economic and political fabric of the late fourteenth and fifteenth centuries AD was entirely different to that of the ninth or tenth centuries, and certainly bore little resemblance to that of the fifth or sixth centuries. It seems inappropriate, then, that vertebrate signatures developed for later medieval England are regularly applied wholesale to the earlier period. This is especially the case since it is known that the meaning ascribed to the procurement and consumption of particular foodstuffs is not static but frequently manipulated in order to maintain social differences (Thomas 2007; van der Veen 2003: 409).

In reality, the ability to label an assemblage as 'high' or 'low' status, or 'Christian' or 'Jewish' is nothing compared to highlighting and elucidating social dynamics. In the absence of documentary evidence, such dynamics can be ascertained archaeologically only by adopting a contextual approach that follows the 'social life' of different foodstuffs (Appadurai 1986; Hamilakis 2000: 57). In order to demonstrate this, we shall turn to the Queen's favourite: the swan.

Life-cycles of Luxury: The Case of the Swan

Perhaps more than any other bird in England, the swan has come to be viewed as a symbol of medieval elite status. The received wisdom is that swans are royal birds and, whilst this is not entirely true, the crown has asserted prerogative rights over swans since the twelfth century. By the mid-thirteenth century the ownership of swans was restricted, as is recorded in Henry de Bracton's *De legibus et consuetudinibus Angliae* (cited in MacGregor 1996: 40). Regulations pertaining to swan ownership are likely, however, to antecede this earliest record, and zooarchaeological evidence affords some insight into the development of the tradition. Figure 8.4 shows that, between the fifth and ninth century, swans were best represented on secular elite settlements. That said, swan remains have also been found on other site types, in particular 'urban' settlements. Their presence on these trading sites is of some significance because these are settlements generally viewed as socially homogeneous and lacking the trappings of socio-economic status: most show a dearth of 'prestige' objects and the diet of their occupants was generally basic and unvaried (O'Connor 2001). The very presence of swan remains on these sites would suggest that this taxon had not yet become the preserve of the elite.

Between the mid-ninth and mid-eleventh centuries, patterns of swan representation appear to have changed: they are absent from rural sites, present on

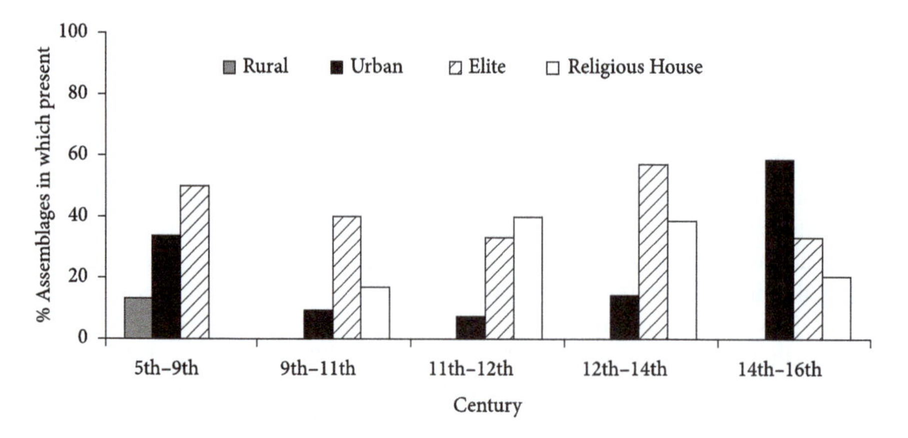

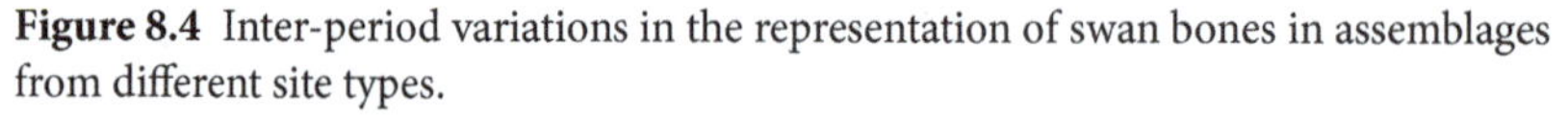
Figure 8.4 Inter-period variations in the representation of swan bones in assemblages from different site types.

Source Sykes (2005).

a small number of urban settlements but more widespread amongst elite assemblages. This suggests that the Late Anglo-Saxon shift in swan representation not only reflects a real change in consumption practices but may also chart the origins of the legal restrictions surrounding swan ownership.

Whilst the swan appears to have become associated with the elite during the ninth to eleventh centuries, the eleventh to twelfth centuries saw swans become better represented in assemblages from religious houses. The reason for this rise in ecclesiastical swan consumption is perhaps provided by the twelfth-century author Alexander Neckham, for whom the swan was an icon of Christian virtue: he saw the transformation of the grey cygnet into the white swan as symbolizing the transformation of the repentant sinner (Rowland 1978: 172). Returning to the concept of humours (pages 18–20), it seems possible that swans were deliberately chosen for their virtuous character. Levels of swan consumption amongst ecclesiastics remained high between the twelfth and fourteenth centuries, the point at which swans begin to be depicted regularly in manuscripts (Yapp 1981: 24); however, this period also saw a peak in the association between swans and elite sites (Figure 8.4).

That the moment of maximum aristocratic swan consumption coincides with the so-called 'Age of Chivalry' is unsurprising. The swan was one of the most important heraldic bird symbols with many noble families, in particular those from the Midlands, claiming descent from the legendary Swan Knight of the medieval romances (Klingender 1971: 460; Coss 2003: 61–5). The integration of its consumption with the rituals of chivalry is perhaps best exemplified by the famous Feast of the Swan, held at Whitsuntide 1306, when three of Edward III's sons were knighted along with almost three hundred others (Coss 2003: 61–5).

As elite demand for swan increased during the mid-twelfth to mid-fourteenth centuries, birds began to be managed upon estates, even within private swanneries. According to historical evidence, swan management increased considerably after the Black Death (Stone 2006). This is supported by the zooarchaeological data, which suggest an overall increase in swan representation: the percentage of assemblages in which they are represented rises from 29 to 36 per cent. What the data also suggest, however, is that the sections of society that were consuming swan changed in this period, with swan becoming less frequent on elite sites but more abundant in urban assemblages (Figure 8.4). It would seem that, as well as raising birds for the rural aristocracy, swanneries sold much of their stock to the urban population. Extant pricelists from urban poulterers and cookshops suggest that swans were always the most expensive birds (Wilson 2003: 118), their cost putting them beyond the reach of the masses. Because of this it may be assumed that many of the specimens recovered from urban contexts of this date derive from the meals of the urban elite. In other cases, however, swans may have been purchased and consumed by the newly emerging middle classes who were gradually adopting aristocratic practices. That this was both taking place and discouraged by the upper classes is clear from the 1482/3 'Act of Swans', which was enacted to dissuade the keeping and consumption of swan by 'persons of little reputation' (MacGregor 1996: 45). This law was clearly ineffective and, as swan consumption became less exclusive, the aristocracy began to scale down their swanneries and reduce their consumption of swan (Stone 2006: 159). Of course, once the elite stopped eating swans, the lower social echelons soon did likewise and from that point these wild birds have slipped from the English menu altogether.

Today, every so often, an episode of swan consumption will hit the newspaper headlines, as in 2006 when a hungry gentleman was arrested for trying to eat a swan. The district judge sentenced the man to 57 days imprisonment stating 'You killed the swan, it was a cruel and reprehensible act. It is a taboo action' (*Daily Mail* 23 November 2006). How times have changed, and how splendid an example of the potency of meat preferences and taboos.

Summary

Given that the majority of zooarchaeological remains derive from animals that were eaten by people, zooarchaeological analyses of food culture have traditionally been rather limited. More emphasis has been placed on calculating the relative dietary contribution made by different animals with far less attention given to elucidating the meanings and dynamics of meat preferences, avoidances and distribution methods.

That zooarchaeologists are only nibbling at the edges of these issues is unsurprising because, essentially, meat is culture, and culture is complex. To reconstruct and interpret the dynamics of meat avoidance and consumption in the archaeological record requires engagement with literature from across the social sciences: anthropology, sociology, psychology and even neuroscience. Furthermore, studies in these fields make clear that we will not achieve the answers to our questions by focusing on the bones alone; instead we must draw upon all available sources of evidence, be they from history, iconography, linguistics, artefacts or scientific analyses.

There are very few individuals who are in a position to weave together such a wide variety of different strands of evidence (Cool 2006 is a notable exception) and, in the main, detailed understandings will be achieved only through the collaboration of researchers from different disciplinary bases. As individuals, however, there are things that we can all strive to do to help move research and discussion forward. First, we should curb the temptation to project our own perceptions of value and taste onto the past: we may not eat horse, fox, insects or swan but that is us, and may not be true of the people that we study. Second, we should avoid projecting ideas about one period, place or context onto another where it may not be appropriate; for instance the application of status criteria from the fifteenth century back onto the fifth century – both centuries may belong to the overarching 'medieval' period but tastes are unlikely to have endured over 1,000 years.

This chapter has sought to highlight that the nature and meaning of meat preferences, avoidances and distribution methods are shifting and situational. Meats that are considered appropriate or powerful in one period and place or for one group, individual or life-stage may be viewed in opposite terms in other circumstances. Similarly whilst meat distribution and consumption can be used to bind groups together and define social position, it can equally be used to subvert the same processes. To identify these nuanced dynamics in the archaeological records requires a bottom-up approach, based fundamentally on the evidence at the level of the context or individual and building up to larger scale cultural interpretations. Such an approach eventually enables temporal trends to be detected. These trends provide not only vital insights into past cultures but also allow us to reflect upon the present, bringing into relief how far removed our culture is from those that we study. The comparison does not always show the present day in a positive light, an issue that will be considered in further detail in the final chapter.

Chapter 9

THE CONCLUSION. ANIMALS AND IDEOLOGY: PAST, PRESENT AND FUTURE

Zooarchaeology is beginning to excite me. In researching and writing this book I have become increasingly convinced about the importance of our discipline. The study of archaeological animals – be they in the form of physical remains, artefacts, images, textual representations or landscapes – is the study of humanity and the human mind. The assemblages we analyse contain information far beyond human economics and diet – they detail not only what people did in the past but also how they viewed their world and considered their position within it. These issues are the very essence of archaeology; they are what I joined the profession to learn about and, after a 20-year hiatus, they are what I want to spend the rest of my career investigating in more detail.

In honesty, integrated analyses of human–animal relationships would have been largely unthinkable twenty years ago. The last two decades have witnessed a revolution in data accessibility, courtesy of the internet, and ground-breaking scientific techniques (e.g. isotope, lipid and DNA analysis) are starting to provide ever-greater resolution to the questions asked by zooarchaeologists. In turn, the tireless efforts of under-appreciated zooarchaeologists have generated excellent and increasingly comparable datasets that can, for the first time, be synthesized to consider how human–animal relationships have changed through time. In this final chapter, I wish to do precisely this and examine the zooarchaeological record over the *longue durée.*

If it is accepted that human–animal interactions are the 'mirrors and windows' to human ideology, it should be possible to bring together all the different English datasets considered in this book and hold them up as a reflection of worldviews over the last 10,000 years. Furthermore, pondering this deep-time evidence allows us to reflect upon our position in the grand scheme of things and contemplate what direction we – the zooarchaeological profession and modern human culture more generally – might wish to take in the future.

Summarizing and Integrating the Data: The Limitations and Caveats

Whilst (zoo)archaeology has reached the point where it is now possible to weave together different datasets and gain a continuous perspective on the past, this kind of analysis is only *just* possible and it is important to recognize that the fabric produced from this weaving is still full of holes (think cheesecloth rather than Harris Tweed). Nevertheless, rather than being paralysed by the fear of inadequate information, I feel it is better to try to do something with our data, if only so that others can develop or contradict the patterns and interpretations that are suggested. The important thing is to be honest about the limitations of the evidence, and so the next few paragraphs are devoted to explaining why you should not believe the graphs that are presented in Figure 9.1.

Even without looking at them in detail, the more statistically savvy readers will instantly pick up on their dubious nature. To quote the current president of the International Council of Archaeozoology, Prof. Terry O'Connor, 'by showing lines linking the point-values for each major period, the graphs imply the range of intermediate values, when in fact we have no idea what the values would be between the plotted points: these graphs need to be redrawn'. The criticism is entirely correct and I probably should have redrawn the graphs. But I have not redrawn them. So you should view them more as illustrative impressions of evidence rather than reliable data. It is also important that you recognize that in this chapter I am flying kites, floating ideas: I expect you to think about my interpretations and challenge them, rather than accept them.

Figure 9.1 shows the different temporal trends exhibited by the various sets of metrical data that have been examined in the previous chapters, with a few additional datasets included. For instance Figure 9.1a summarizes the inter-period variations in average $\delta^{13}C$ of human bone collagen, which I am suggesting can be used as a proxy for levels of marine food consumption. These $\delta^{13}C$ data are a simplified version of those presented in Figure 3.4 and the sources of the data are the same. Likewise, Figure 9.1b simplifies the percentage of wild animals in zooarchaeological assemblages shown in Figure 3.1 – the limitations of both these datasets are discussed in Chapter 3.

Figure 9.1c presents information that has not hitherto been considered in any detail: the changing size of cattle. Here the greatest length of cattle metacarpals are shown (for the Mesolithic period, aurochsen metacarpals are used), the idea being that they provide a general indication of cattle size change through time. I stress the word 'general' because one skeletal element does not make an animal: a single body-part can show size variations entirely independent of overall body size. Nevertheless, the metacarpals are the best we have to work with; they preserve well and are recovered frequently, rendering their sample size sufficient for inter-period analysis. The data for this graph have been drawn from multiple sources: for the aurochs the measurements come from personally collected data plus those presented in Everton (1975); the Neolithic data are from Grigson (1999), the Bronze Age data are derived primarily from Legge (1992) and the remainder have been downloaded from the invaluable on-line resource, the Animal Bone Metrical Archive Project (see Serjeantson 2005).

Figures 9.1d and 9.1e show the inter-period variations in the number of vertebrate extinctions and introductions: the data for these graphs are taken from O'Connor and Sykes (2010) and Lever (2009). Figure 9.1f summarizes variations

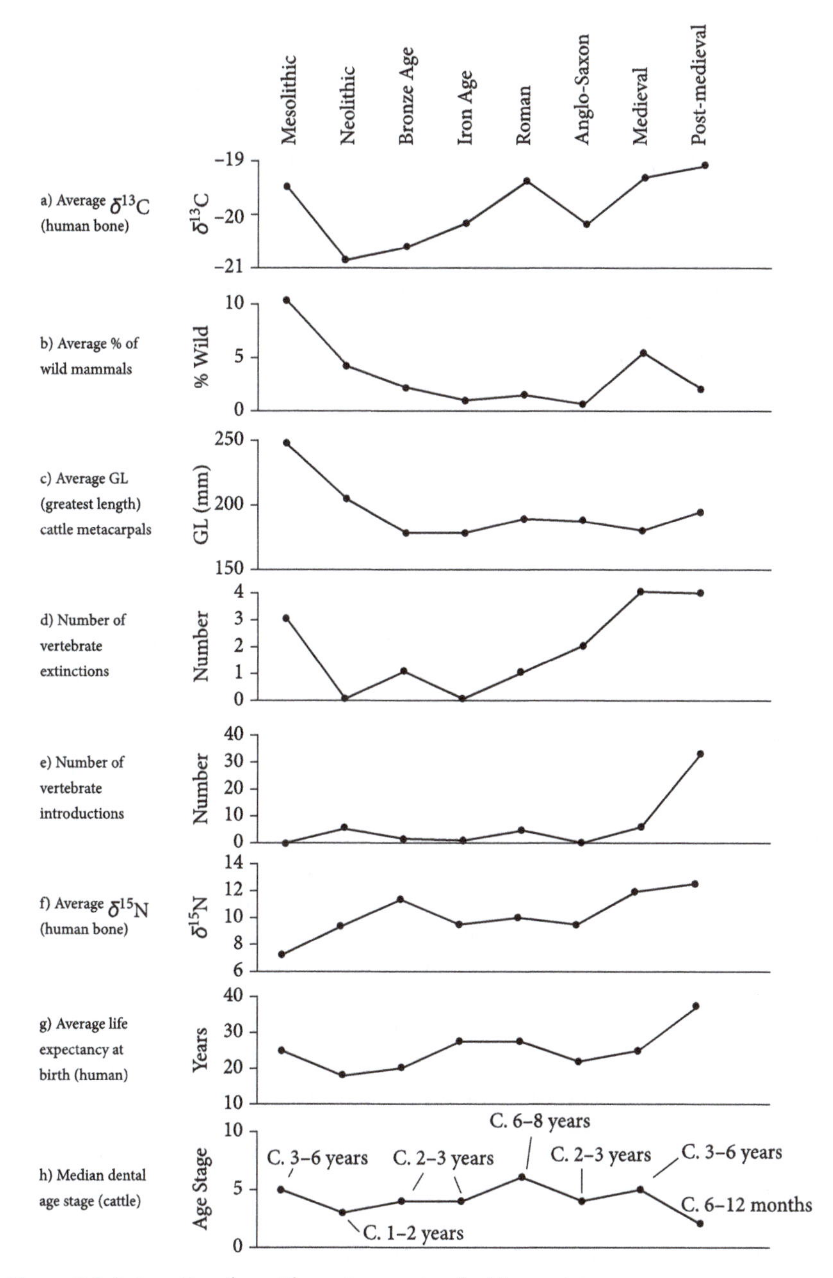

Figure 9.1 Integrating the evidence to examine the bigger picture.

in average $\delta^{15}N$ of human bone collagen, which I am suggesting can be used as a proxy for levels of animal protein consumption. There are problems with this proposal as nitrogen values can be influenced by many different factors (see Hedges and Reynard 2007); however, with this caveat in place, they are presented here, the data derived from the same sources as those in Figure 3.2. Perhaps the most problematic data are those shown in Figure 9.1g, which provides a generalized suggestion of how human life expectancy at birth has changed through time. Calculations of human life expectancy at birth are fraught with difficulties, especially for Prehistoric populations (Chamberlain 2006). Figure 9.1g has been cobbled together from the best estimates presented by period specialists (Benedictow 2004; Bocquet-Appel 2011a, 2011b; Butler 2008; Harding 2000; Wells 2011). The point to note is that, for each period, we must expect that lifespans were, for some individuals, very much shorter, or much longer, than the figure presented in the graph (Chamberlain 2006: 53). The same caveat applies to Figure 9.1h, which shows the inter-period variations in the median age stage achieved by cattle, which I am using as a proxy for all domestic animals. This graph has been calculated from raw mandibular tooth eruption and wear data, extracted from several synthetic studies (Allen 2010; Serjeantson 2011; Sykes 2006b) and from other personally collected data.

Human–Animal Interactions in Transition

When viewed together, the graphs presented in Figure 9.1 make interesting reading. They reiterate the points that have been made in the preceding chapters but there are some surprises, which I had not anticipated before I brought all the data together. For instance, the evidence for the Mesolithic period confirms that the diet of hunter-gatherers contained a high proportion of protein derived from marine sources, as is suggested by Figure 9.1a, and indicates that wild mammals dominated the faunal assemblages (Figure 9.1b). Throughout this book I have given a rather idealized impression of the Mesolithic period, suggesting that the hunter-gatherers of the period lived in 'harmony' with the environment, their relationship with their surroundings being familial and based on trust. The proposition has been that human–animal interactions were respectful with people taking no more than they need, and the evidence from Figure 9.1f indicates that protein consumption was comparatively low in the Mesolithic period, suggesting that a higher proportion of the diet may have been plant-based. With these Garden of Eden-like interpretations in mind, it is interesting to note that the Mesolithic period saw one of the highest rates of vertebrate extinction in British history, with the demise of the elk, reindeer and horse (Figure 9.1d). To be sure, some of these extinctions can be related to climatic change but pressure from hunting cannot have helped; perhaps the Mesolithic hunters were not as respectful as we might have anticipated. Certainly for the continent, Conneller (2011: 363) has identified evidence for episodes of unnecessary overkill, and many sites appear to suggest an emphasis on the hunting of female and juvenile animals – demographic groups that are normally avoided in sustainable hunting strategies (Leduc 2012; Prummel and Niekus 2011; Richter 1982).

The Mesolithic evidence from England is less abundant and most of the aurochsen remains recovered from hunter-gatherer sites tend to be from old or elderly animals (figure 9.1h, see also Legge and Rowley-Conwy 1988). Considering that, in the Mesolithic, the human life expectancy at birth was about 25 years of age, it is possible to estimate that the average life expectancy of an aurochs (3 to 6 years) equated to approximately 20 per cent of a human's life. This human-to-cattle life expectancy ratio is an interesting measure, particularly when viewed over the long-term; however I should stress that the President of ICAZ has begged me not to present these comparisons as he believes them to be statistically invalid and therefore, potentially, misleading. Again, I am ignoring his good advice but you should keep his criticism in mind for the rest of the chapter.

For the Neolithic period, the ratio of human-to-cattle life expectancy ratio changes quite significantly, with the median cattle age dropping (to 1 to 2 years), just 11 per cent of a Neolithic human's life expectancy. Whilst this relative reduction in cattle age could suggest the kind of human domination over animals suggested by Ingold (1994), whereby cattle were exploited simply for the benefit of people, the average age for both cattle *and* humans drops in this period (the latter to 18 years of age). There are several reasons for this – some of which relate to shifts in burial practice – but the shifts in human and cattle age may not be entirely unconnected; indeed, they may both be linked to dairying. As set out in Chapter 2, it is clear from pottery residue analysis that domestic cattle arrived in Britain bringing with them a fully fledged dairying 'economy' (Copley *et al.* 2003, 2005; Craig *et al.* 2005). This would have necessitated the slaughter of juvenile animals (hence the decline in cattle age); however, the effect of milk consumption on the lactose-intolerant human body may equally explain the drop in human life expectancies at birth. All human children demonstrate high lactase expression at birth but this declines quickly after weaning, particularly in lactose non-persistent individuals (Ingram *et al.* 2009: 579; Leonardi *et al.* 2012). It is widely recognized that Neolithic children were weaned earlier than their Mesolithic counterparts, perhaps within 2 years (e.g. Howcroft 2012; Richards *et al.* 2003). Weaned children who were lactase non-persistent would have become more susceptible to mortality from infectious diseases such as diarrhoea, since drinking milk would have exacerbated their condition (Ingram *et al.* 2009: 588). Bocquet-Appel (2011a, 2011b: 560) has suggested that mortality from diarrhoea was responsible for the dramatic Mesolithic to Neolithic increase in the proportion of 5- to 19-year-old individuals in cemeteries. If these mortalities were largely lactase non-persistent individuals, the same group may also have been responsible for positive selection for lactase persistence (Leonardi *et al.* 2012: 93).

For all the demographic change, dairy products appear to have brought dietary benefits to the human population, as is perhaps indicated by Figure 9.1f, which shows a slight rise in the average $\delta^{15}N$ levels of human bone collagen. It would also seem that these animal proteins came almost entirely from domestic sources, since Figure 9.1a and 9.1b indicate a rapid shift away from the exploitation of wild resources from marine and terrestrial ecosystems (see Chapter 3). This new focus on domestic animals may have given wild animal populations time to recover from Mesolithic hunting and there is no clear evidence that any vertebrate species went extinct during the Neolithic period. To some extent, this situation contrasts with the recent suggestions that the arrival of farming represents humanity's downfall – that everything that is bad in the modern world can be traced to the transition from a hunter-gatherer to farming way of life (Wickham-Jones 2010). Instead, the graphs presented in Figure 9.1, together with the evidence for Neolithic mortuary rituals, which frequently involved the co-mingling of human and animal remains, suggest a situation of greater human–animal intimacies and of greater respect for the natural world than has been seen in any period before or after. This is also something suggested by genetic studies, which highlight the explosion of animal coat colours that accompanied domestication: as I argued in Chapter 2, variation in coat colour both permits and develops from closeness in human–animal relations.

A better case for human dominion over animals can, perhaps, be made based on the data for the Bronze Age, when the average size of cattle metacarpals drops substantially (Figure 9.1c), potentially suggesting that people were curbing the ability of animals to thrive in the way that they might outside of human influence. That said, the median age stage reached by domestic cattle increases in the Bronze Age to 2 to 3 years, representing 15 per cent of a human life expectancy. This shift in cattle age structure could indicate a reduced reliance on dairying and a greater emphasis on plough animals but, regardless of economic motivation, it is interesting to note that the shift in age structure is again reflected by the human data: Figure 9.1g shows that human life expectancy at birth rose to 20 years of age. It is possible that human longevity was facilitated by greater access to dietary protein, Figure 9.1f suggesting that the average $\delta^{15}N$ levels of human bone collagen rose during the Bronze Age. However, the reliability of these ageing and isotope data are very questionable because the sample size on which the Bronze Age plot has been calculated is very small and it is clear that inhumation was reserved for specific people, notably adult men of high status (Bradley and Fraser 2010). Certainly it was in the Bronze Age that elite individuals, particularly male 'warriors', become more visible in the archaeological record. As set out in Chapter 3, the identity and power of these individuals was most probably expressed and negotiated through warfare and hunting, and it may be no coincidence that the British aurochs was driven to extinction at this point. Beyond the hunting of the aurochs, the exploitation of wild resources, both terrestrial and marine, appears to have been limited, as is indicated by both Figure 9.1a and 9.1b. Instead, greater attention was given to exciting new animal arrivals; specifically the horse, archaeological representations of which begin in the Bronze Age, becoming particularly apparent in the Iron Age.

The cultural significance of the 'exotic' horse in the Iron Age is reflected by the art and archaeology of the period, and it would seem that horse-man deities joined the pantheon of human–animal hybrid gods that were found across much of contemporary Europe and beyond. In England, international contact appears to have declined at the end of the Bronze Age and the Iron Age saw the emergence of a more insular lifestyle focused on the domestic. The arrival of woolly sheep may have radically altered human–animal interactions and there is evidence that sheep take a more central role in the zooarchaeological record of the Iron Age (Serjeantson

2007); however the evidence relating to cattle (Figure 9.1c and 9.1h) suggests that human–cattle relations continued as they had in the Bronze Age. Wild animals also continued to be left largely to their own devices and, if not already in existence, cultural taboos surrounding the 'wilderness' may have developed or strengthened, evidenced by votive practices and the excarnation funerary rites of the Iron Age (Chapter 3). With the exception of the chicken and, perhaps, the domestic cat, no new animals arrived in Iron Age Britain but neither did any large species go extinct.

This situation changed dramatically into the Roman period. For although the number of animal extirpations is surprisingly low (the pelican was the only species that departed Britain's shores in the Roman period), Figure 9.1e shows a peak in the number of animal introductions, which can be seen more clearly in Figure 4.1. In Chapter 3, I argued that the incorporation of Britain into the Roman Empire changed the conceptual boundaries of the known world, rendering it permissible to capture and eat the wild resources previously considered sacred. This is indicated by the slight increase in the average percentage of wild animals seen in Roman assemblages (Figure 9.1b) and the more dramatic rise in the average $\delta^{13}C$ levels of human bone collagen, which suggest that marine resources were becoming a frequent dietary component (Figure 9.1a). The average $\delta^{15}N$ levels (Figure 9.1f) also demonstrate a slight increase in the Roman period and, if it is accepted that these isotope data provide a reliable indication of diet, it may suggest that levels of protein and nutrition in general improved during the Roman period. The economic and landscape evidence suggest a Roman concern with productivity, with estates being farmed for maximum human benefit and the 'wilderness' being put under the plough. The central importance of cattle for traction and manure goes some way to account for the ageing data shown in Figure 9.1h, which shows the very old age that cattle achieved before being sent to market so that their carcasses could be swiftly and efficiently processed into standardized cuts by professional urban butchers.

The need for large powerful animals may also explain the well-known trend for Roman cattle to have a larger carcass size than those of the Iron Age (Figure 9.1c and Albarella *et al.* 2008). However, economics alone do not appear adequate to account for the size and surprisingly old age of the Roman cattle: the median stage of Roman cattle was approximately 6 to 8 years, equivalent to 25 per cent of a contemporary human's life expectancy. For me, the exceptional age of these cattle is shocking, requiring me to re-think my perception of Roman attitudes to the natural world. Throughout this book I have concluded the Romans to have been human supremacists (indicated by the shift away from human–animal hybrid deities and excarnation – see Chapter 3), who saw it as their spiritual duty to dominate nature and bring it to order. I am still of this opinion, and certainly the Roman Empire spelled ecological disaster for many of the areas it encompassed (Hughes 2003); however, the cattle ageing data suggest to me these Roman animals may have had an importance beyond that of mere meat and traction: could they, in life, have symbolized the very core of Roman identity? Installation of large numbers of large cattle within the Romano-British landscape would have transformed its look, sound and smell: a cultural conquest of the senses. MacKinnon (2010b: 56) alludes to the significance of cattle within Roman culture but, based on the ageing data shown in Figure 9.1h, it is a point that probably deserves to be made more forcefully. Particularly since, rather than the kind of despotic exploitation usually attributed to the Romans, the relationship that the (rural?) population had with the majority of their cattle must have been close and long-term: the scenario I now envisage is one more akin to the traditional cattle culture of the Siberian Saka (Crate 2008).

This cattle culture endured initially into the early Anglo-Saxon period but, with the withdrawal of the Roman Empire from Britain, all the graphs in Figure 9.1 show the same trend: essentially a reversion to the Iron Age lifestyle and worldview. The wilderness, it would seem, became once again a sacred or a scary place from which foodstuffs were not acquired (Figure 9.1a and 9.1b). The average $\delta^{15}N$ level falls, as does human life expectancy at birth, the two possibly being connected. In contrast to the previous period of animal introductions, Anglo-Saxon England saw no new arrivals. On the other hand, rates of animal extinction seem surprisingly high, especially given the low percentages of wild animals in the zooarchaeological assemblages of this date. The lynx and the bear both appear to have become locally extinct in the Anglo-Saxon period; and the fact that Anglo-Saxon funerary rites involved bear skins (Bond and Worley 2006) – and iconographic evidence suggests a fascination with dangerous animals (Pluskowski 2010) – it may be that the Anglo-Saxon population were to blame for the extirpation of these species.

Collectively, the graphs in Figure 9.1 suggest that the medieval period represents the beginnings of genuine human dominion, with people exploiting the environment for their own benefit. That humans did benefit is clear from Figure 9.1f, where it can be seen that the medieval period saw a substantial rise in average $\delta^{15}N$ levels and marks the start of expanding human life expectancies (Figure 9.1g). As ever, this shift in human life-span is apparently mirrored by their livestock, with median cattle ages almost doubling from approximately 2 to 3 years to 3 to 6 years. Generally this increase in cattle age has been attributed to a renewed emphasis on agriculture, with cattle being utilized increasingly for traction, as they had been in the Roman period (Sykes 2006b). However, by contrast to the Roman period, there is less evidence that people valued their cattle: Figure 9.1c suggests that animal size dropped to its lowest point in human history – a trend that has been linked to nutrient deficiency from poor grazing (Sykes 2007b: 51) – and there is little in the historical record to suggest that cultural identities were forged through human–cattle relationships in this period. Indeed, the Christian monotheistic worldview would have taken a dim view of such relationships (Chapter 7).

The good Christians of the medieval period appear to have taken up the challenge of widespread exploitation. They ate large quantities of marine animals (Figure 9.1a) that, according to humoural principles, were deemed a fitting food for the pious. And they hunted and ate large quantities of wild land mammals (Figure 9.1b) as a symbol of their cosmological supremacy. The consequences of these actions can be seen in Figure 9.1d, which highlights the rise in local extinctions brought by the medieval period. The wolf was the most iconic casualty but many other species, such as the red deer, roe deer and the beaver, were also brought close to extirpation. In their place came imported exotic creatures, slightly more than occurred in the Roman period but far fewer than were to follow in the succeeding centuries (9.1e and Chapter 3).

The post-medieval period represents a gear-change in human–animal relationships. Figure 9.1e shows the hockey stick-shape curve of animal introductions, the dramatic rise reflecting the unprecedented global trade of the time (Chapter 4). Whereas other periods that witnessed high levels of international trade (e.g. the Bronze Age, Roman and medieval periods) were also typified by hunting cultures, the post-medieval period saw rates of wild animal exploitation decline (Figure 9.1b). To some extent this is because ungulate hunting was falling out of fashion, but the fact that few wild ungulates were available may have influenced the decision: Figure 9.1d shows that rates of animal extinctions continued at the rate seen in the medieval period. Exploitation of marine resources, however, appears to have continued, with Figure 9.1a suggesting that average $\delta^{13}C$ values for human bone collagen reached their highest levels in human history during the post-medieval period. The same is true for average $\delta^{15}N$ levels (Figure 9.1f) and, again, this graph is correlated with an increase in human life expectancy (Figure 9.1g).

Throughout most of human history, human life expectancy and median cattle age appear to have risen and fallen in tandem, Figures 9.1g and 9.1h demonstrating closely paralleled trends that suggest a level of human–cattle partnership. The one period that deviates from this pattern of unison is the post-medieval when, for the first time, the two graphs bifurcate: human life expectancy reaches an all-time high of 37 years, whereas median cattle age reaches an all-time low of less than 1 year, equivalent to just 3 per cent of the post-medieval human life-expectancy at birth. To me, this shift in the human-to-cattle age ratio is the most eloquent expression of the fracturing of human–animal relationships, when animals moved from being part of a community to being nothing more than commodities. This is the period that, as I argued in Chapter 2, saw the emergence of livestock 'breeds' – a cultural concept that transformed livestock herds previously composed of individual animals of varied size, conformation and colour, into homogenized troops of economic units, bred according to human-designed blueprints of breed 'standard'.

When viewed over the *longue durée* the post-medieval situation leaves me feeling a little queasy. First, because the post-medieval period was the 'Age of Enlightenment', the time of scientific revolution when people thought they had it so right but, according to the data presented in Figure 9.1, they would appear to have been mistaken. Second, because I have a nasty feeling that I know what comes next. For reasons set out by Thomas (2009) we lack good archaeological data for the time since the post-medieval period but I think we are safe to speculate about how we arrived at the present day.

I believe that we could be fairly confident that early modern and present day rates of meat consumption would surpass those of the post-medieval period, something that is difficult to reconstruct today from isotope studies due to the complexities of the modern international food chain. My suspicion is that modern rates of wild mammal consumption would be very low, perhaps comparable with levels of the Iron Age, with whom we share a reverence for the 'wilderness' (Chapter 3). Our current cultural taboo over wild animal consumption comes at a time, however, when there are no longer top predators – wolves, bears and the lynx – to keep the country's deer numbers in balance. Recent research suggests that deer numbers are at an all-time-high, yet there is no UK market for wild venison, so the majority of meat produced by English deer-stalkers is exported to the continent (Riminton personal communication). Paradoxically, if you did wish to buy some venison from an English supermarket, the chances are it would be imported from New Zealand (Riminton personal communication).

Whilst the hunting of abundant free-range wild deer is widely considered 'cruel', the English public rely increasingly on food derived from over-fished seas or on the meat provided by a limited number of farm animals, slaughtered at ever younger ages. Comparatively, the average slaughter age of cattle is today similar to that seen in the post-medieval period, the majority of animals are culled at under a year of age. The major difference is that humans now live much longer (in Western culture life expectancy at birth is 80 years), which means that, today, an average cattle life represents just 1 per cent of a human life-span, quite a contrast with the 25 per cent ratio of the Roman period. Things are much worse for chickens. In Chapter 7 it was seen that in seventh-century Vienna many of the chickens that were buried with humans lived to several years of age and Doherty (2013) has argued that the majority of Roman cockerels may have lived in excess of 3 years (14 per cent of Roman life expectancies at birth). Compare this with the present-day situation. The McDonalds website provides very useful information about the slaughter ages of their animals. They state that their chickens – which are reared for meat, not eggs – are culled between 35–42 days, marginally older than the lifespan of a house fly (25 days) but significantly less than that of a mosquito (100 days) or bee (120 days) and representing a mere 0.14 per cent of a modern human's life expectancy (Fredericks 2010).

The queasiness I felt a few paragraphs ago has now turned to horror at how far human–animal relationships have been transformed into one-sided exploitation of animals and the environment. In 1973, E. F. Schumacher reminded us that we talk about humanity's 'battle with nature' all the while forgetting that, if we were to win, we would find ourselves on the losing side (Schumacher 1973: 3). Examining the evidence presented in Figure 9.1, I fear we could be close to both winning and losing our battle. However, Figure 9.1 has made me realize that zooarchaeologists are uniquely placed to consider the human past and, by reflecting upon our history, we are one of very few disciplines with the privilege of a deep-time overview of human–nature relationships. As with all privileges, this comes with some responsibility and it is perhaps time for zooarchaeologists to step up to the mark and shout forcefully about the lessons that our work on ancient societies has for the future. This is a responsibility that, with very few exceptions (Grayson 2001; Lauwerier and Plug 2005) we have largely shirked, and I am particularly guilty of this.

Take Chapter 4 of this book, for instance: it overlooks entirely, albeit deliberately, the importance of zooarchaeological evidence to present-day issues of biodiversity. This is despite the fact that the zooarchaeological record provides empirical data regarding the ecological and cultural impact of species introductions/extinctions, the importance of which for modelling sustainable practice in the future is becoming widely recognized (e.g. Erlandson and Rick 2010; Lambert and Rotherham 2011; Scandura *et al.* 2011; Sykes and Putman in press). Indeed, the International Union for the Conservation of Nature (IUCN) guidelines for reintroductions specify that the factors responsible for a species' extinction must be identified before a reintroduction can be considered: it would seem that zooarchaeology is the very discipline that can help with this task and contribute to the growing discussions concerning 're-wilding' (e.g. Hetherington *et al.* 2006).

The relevance of zooarchaeological evidence for modern wildlife management has not been lost on the likes of George Monbiot, whose highly praised 2013 book *Feral* is littered with evidence from archaeological animal remains. I read Monbiot's *Feral* with a mixture of both interest and annoyance: it is a thought-provoking book but why was it not written by a zooarchaeologist? We work hard to produce our data yet it seems that we are the very people most blind to their significance. I wonder if it is because, collectively, we have started to believe what 'real' archaeologists tell us: that we are a niche specialism with nothing to offer. Occasionally, some zooarchaeologists do try to make the case that others should listen to what we have to say (Maltby 2006) but, generally, this is a conversation had amongst and between zooarchaeologists: we are the equivalent of the bystanders who watch a public assault, whispering to each other that really we should do something (without actually doing anything).

There has never been a better time for zooarchaeology to come to the fore of archaeological research. As I write, the field of Human–Animal Studies is becoming genuinely mainstream, with scholars from different disciplines recognizing the need to study human–animal interactions. Archaeology would benefit greatly by engaging with current debates and zooarchaeologists are the natural brokers for this exchange. However, we have the opportunity not only to bring new perspectives about the past; the growing number of journals, edited volumes and monographs dedicated to the 'Animal Turn' means that zooarchaeologists can also contribute to discussions about the present. Data derived from zooarchaeological research can enhance understanding about the ethics of animal management and exploitation, and can challenge the very foundations of modern day anthropocentrism. That is exciting.

Whether or not zooarchaeologists seize this opportunity remains to be seen but certainly many are trying (e.g. Armstrong Oma 2010; O'Connor 2013a; Overton and Hamilakis 2013). I hope this book makes a small contribution to the movement but, at a personal level, it has done what I needed it to achieve: I was bored when I started writing but, for the first time in a long time, I am looking forward to work tomorrow.

REFERENCES

Abbink, J. (1993), 'Reading the Entrails: Analysis of an African Divination Discourse', *Man*, 28: 705–26.

Abbink, J. (2003), 'Love and Death of Cattle: The Paradox in Suri Attitudes Towards Livestock', *Ethnos*, 68(3): 341–64.

Abdalla, M. (1994), 'Milk in the Rural Culture of Contemporary Assyrians in the Middle East', in P. Lysaght (ed.), *Milk and Milk Products from Medieval to Modern Times*, Edinburgh: Canongate Academic.

Affani, G. (2008), 'Astragalus Bone in Ancient Near East: Ritual Depositions in Iron Age I in Tell Afis', in J. Molist, M. Pérez, I.L. Rubio and S. Martínez (eds), *Proceedings of the 5th International Congress on the Archaeology of the Ancient Near East. Madrid, April 3–8 2006*, Córdoba: Universidad Autónoma de Madrid.

Albarella, U. (1997a), 'Shape Variation of Cattle Metapodials; Age, Sex or Breed? Some examples from Medieval and Post-medieval sites', *Anthropozoologica*, 25–6: 37–47.

Albarella, U. (1997b), 'Size, Power, Wool and Veal: Zooarchaeological Evidence for Late Medieval Innovations', in G. De Boe and F. Verhaeghe (eds), *Environment and Subsistence in Medieval Europe*, Bruges, Belgium: Institute for the Archaeological Heritage of Flanders.

Albarella, U., Johnstone C. and Vickers, K. (2008), 'The Development of Animal Husbandry from the Late Iron Age to the end of the Roman Period: A Case Study from South-East Britain', *Journal of Archaeological Science*, 35: 1828–48.

Albarella, U. and Serjeantson, D. (2002), 'A Passion for Pork: Meat Consumption at the British Late Neolithic Site of Durrington Walls', in P. Miracle and N. Milner (eds), *Consuming Patterns and Patterns of Consumption*. Cambridge: MacDonald Institute.

Albarella, U. and Thomas, R. (2002), 'They Dined on Crane: Bird Consumption, Wildfowling and Status in Medieval England', *Acta Zoologica Cracoviensia*, 45: 23–38.

Aldhouse-Green, M. (2004), *An Archaeology of Images: Iconology and Cosmology in Iron Age Roman Europe*. London: Routledge.

Algaze, G. (2009), *Ancient Mesopotamia at the Dawn of Civilization: The Evolution of an Urban Landscape*. Chicago: University of Chicago Press.

Allen, M. (2010), *Animalscapes and Empires: New Perspectives on the Iron Age/Romano-British Transition*. Unpublished PhD thesis, University of Nottingham.

Allen, M. and Sykes, N. (2011), 'New Animals, New Landscapes and New Worldviews: The Iron Age to Roman Transition and Fishbourne', *Sussex Archaeological Collections*, 149: 7–24.

Almond, R. (2003), *Medieval Hunting*. Stroud: Sutton.

Alvard, M.S. (1995), 'Intraspecific Prey Choice by Amazon Hunters', *Current Anthropology*, 36(5): 789–818.

Alves, R.R.N. and Rosa, I.L. (2005), 'Why Study the Use of Animal Products in Traditional Medicines?', *Journal of Ethnobiology and Ethnomedicine*, 1: 1–5.

Alves, R.R.N., Medeiros, M.F.T., Albuquerque, U.P. and Rosa, I.L. (2013), 'From Past to Present: Medicinal Animals in a Historical Perspective', in R.R.N. Alves and

I.L. Rosa (eds), *Animals in Traditional Folk Medicine: Implications for Conservation*, Berlin: Springer.
Amanzadeh, J., Gitomer, W.J., Zerwekh, J., Preisig, P. A., Moe, O.W., Pak, C.Y. and Levi, M. (2003), 'Effect of High Protein Diet on Stone-Forming Propensity and Bone Loss in Rats', *Kidney International*, 64: 2142–9.
Ambrose, S.H. and Norr, L. (1993), 'Experimental Evidence for the Relationship of the Carbon Isotope Ratios of Whole Diet and Dietary Protein to those of Bone Collagen and Carbonate', in J.B. Lambert and G. Grupe (eds), *Prehistoric Human Bone: Archaeology at the Molecular Level*, New York: Springer.
Amoroso, E.C. and Jewell, P.A. (1963), 'The Exploitation of the Milk-Ejection Reflex by Primitive Peoples', *Man and Cattle*, 18: 126–37.
Anderson, E. N. Jr. (1987), 'Why is Humoral Pathology so Popular?', *Social Science and Medicine*, 25: 331–7.
Anderson, J.K. (1985), *Hunting in the Ancient World*, London: University of California Press.
Anthony, D.W. and Brown, D.R. (2011), 'The Secondary Products Revolution, Horse-Riding, and Mounted Warfare', *Journal of World Prehistory*, 24(2–3): 131–60.
Appadurai, A. (1986), 'Introduction: Commodities and the Politics of Value', in A. Appadurai (ed.), *The Social Lives of Things: Commodities in Cultural Perspective*, Cambridge: Cambridge University Press.
Arbuckle, B.S. (2005), 'Experimental Animal Domestication and its Application to the Study of Animal Domestication in Prehistory'. in J-D. Vigne, J. Peters and D. Helmer (eds), *The First Steps of Animal Domestication: New Archaeozoological Techniques*, Oxford: Oxbow.
Arbuckle, B.S. (2012), 'Animals in the Ancient World', in D.T. Potts (ed.), *A Companion to the Archaeology of the Ancient Near East*, Oxford: Wiley-Blackwell.
Arbuckle B.S., Öztan A. and Gülçur S. (2009), 'The Evolution of Sheep and Goat Husbandry in Central Anatolia', *Anthropozoologica*, 44(1): 129–57.
Arce, J. (2010), 'Roman Imperial Funerals in Effigie', in B.C. Ewald and C.F. Noreña (eds), *The Emperor and Rome: Space, Representation, and Ritual*, Cambridge: Cambridge University Press.
Arikha, N. (2007), *Passions and Tempers: A History of the Humours*, New York: Harper Collins.
Armit, I. (2011), 'Violence and Society in the Deep Human Past', *British Journal of Criminology*, 51: 499–517.
Armstrong Oma, K. (2010), 'Between Trust and Domination: Social Contracts between Humans and Animals', *World Archaeology*, 42(2): 157–85.
Ashby, S.P. (2002), 'The Role of Zooarchaeology in the Interpretation of Socioeconomic Status: A Discussion with Reference to Medieval Europe', *Archaeological Review from Cambridge*, 18: 37–60.
Ashmore, W. and Knapp, A.B. (1999), *Archaeologies of Landscape: Contemporary Perspectives*, Oxford: Blackwell.
Aston, M. and Rowley, T. (1974), *Landscape Archaeology: An Introduction to Fieldwork Techniques on Post-Roman Landscapes*, Newton Abbot, UK: David & Charles.
Aufderheide, A. and Rodriguez-Martin, C. (1998), *Cambridge Encyclopedia of Human Paleopathology*, Cambridge: Cambridge University Press.
Auguet, R. (1972), *Cruelty and Civilization: The Roman Games*, London: Routledge.
Aybes, C. and Yalden, D.W. (1995), 'Place-name Evidence for the Former Distribution and Status of Wolves and Beavers in Britain', *Mammal Review*, 25(4): 201–26.

Bahn, P. (1989), *Bluff Your Way in Archaeology*, Horsham, UK: Ravette.
Bahn, P.G. and Vertut, J. (1988), *Journey Through the Ice Age*, Los Angeles: University of California.
Baker, J.R. and Brothwell, D.R. (1980), *Animal Diseases in Archaeology*, London: Academic Press.
Baker, K. (2011), *Population Genetic History of the British Roe Deer (Capreolus capreolus) and its Implications for Diversity and Fitness*, Unpublished PhD thesis, Durham University, UK.
Balasse, M. (2002), 'Reconstructing Dietary and Environmental History from Enamel Isotopic Analysis: Time Resolution of Intra-tooth Sequential Sampling', *International Journal of Osteoarchaeology*, 12(3): 155–65.
Balasse, M. and Tresset, A. (2002), 'Early Weaning of Neolithic Domestic Cattle (Bercy, France) Revealed by Intra-tooth Variation in Nitrogen Isotope Ratios', *Journal of Archaeological Science*, 29(8): 853–9.
Barber, E.J.W. (1991), *Prehistoric Textiles: The Development of Cloth in the Neolithic and Bronze Ages with Special Reference to the Aegean*. Princeton, NJ: Princeton University Press.
Barber, E.J. (1995), *Women's Work: The first 20,000 Years: Women, Cloth, and Society in Early Times*, New York: Norton & Company.
Barker, G. (2006), *The Agricultural Revolution in Prehistory: Why Did Foragers Become Farmers?* Oxford: Oxford University Press.
Barnes, M.P. (2012), *Runes: A Handbook*, Woodbridge, UK: Boydell Press.
Barrett, J.H., Locker, A.M. and Roberts, C.M. (2004), '"Dark Age Economics" Revisited: The English Fish Bone Evidence AD 600–1600', *Antiquity*, 78: 618–36.
Barrett, J.H. and Richards, M.P. (2004), 'Identity, Gender, Religion and Economy: New Isotope and Radiocarbon Evidence for Marine Resource Intensification in Early Historic Orkney, Scotland, UK', *European Journal of Archaeology*, 7(3): 249–71.
Barringer, J.M. (2001), *The Hunt in Ancient Greece*. Baltimore, MD: JHU Press.
Barton, L., Newsome, S.D., Chen, F.H., Wang, H., Guilderson, T.P. and Bettinger, R.L. (2009), 'Agricultural Origins and the Isotopic Identity of Domestication in Northern China', *Proceedings of the National Academy of Sciences*, 106(14): 5523–8.
Bartosiewicz, L. (2003), 'There's Something Rotten in the State . . .: Bad Smells in Antiquity', *European Journal of Archaeology*, 6(2): 175–95.
Bartosiewicz, L. (2008), 'Environmental Stress in Early Domestic Sheep', in Z. Miklíková and R. Thomas (eds), *Current Research in Animal Palaeopathology: Proceedings of the Second Animal Palaeopathology Working Group Conference*, (British Archaeological Reports International Series S1844), Oxford: Archaeopress.
Bartosiewicz, L. (2012), '"Stone Dead": Dogs in a Medieval Sacral Space', in A. Pluskowski (ed.), *The Ritual Killing and Burial of Animals: European Perspectives*, Oxford: Oxbow.
Baxter, I.L. (2006), 'A Dwarf Hound Skeleton from a Romano-British Grave at York Road, Leicester, England, UK, with a Discussion of Other Roman Small Dog Types and Speculation Regarding their Respective Aeitiologies', in L.M. Snyder and E.A. Moore (eds), *Dogs and People in Social, Working Economic or Symbolic Interaction*, Oxford: Oxbow.
Beagon, M. (1996), 'Nature and Views of Her Landscapes in Pliny the Elder', in J. Salmon and G. Shipley (eds), *Human Landscapes in Classical Antiquity: Environment and Culture*, London: Routledge.
Beal, R.H. (2002), 'Hittite Oracles', in L. Ciraolo and J. Seidel (eds), *Magic and Divination in the Ancient World*, Leiden: Brill.

Beckett, J.V. (1990), *The Agricultural Revolution*, Oxford: Blackwell.
Beeton, I. (1861), *The Book of Household Management*, London: Octavo.
Beetz, A.M. and Podberscek, A.L. (2005), *Bestiality and Zoophilia: Sexual Relations with Animals*, New York: Purdue University Press.
Beirne, P. (1994), 'The Law is an Ass: Reading E.P. Evans' the Medieval Prosecution and Capital Punishment of Animals', *Society and Animals*, 2: 27–46.
Beirne, P. (2000), 'Rethinking Bestiality: Towards a Concept of Interspecies Sexual Assault', in A.L. Podberscek, E.S. Paul and J.A. Serpell (eds), *Companion Animals and Us: Exploring the Relationship Between People and Pets*, Cambridge: Cambridge University Press.
Beirne, P. (2013), 'Hogarth's Animals', *Journal of Animal Ethics*, *3*(2), 133–62.
Bendrey, R. (2007), 'New Methods for the Identification of Evidence for Bitting on Horse Remains from Archaeological Sites', *Journal of Archaeological Science*, 34(7): 1036–50.
Bendrey, R. (2010), 'The Horse', in T. O'Connor and N. Sykes (eds), *Extinctions and Invasions: A Social History of British Fauna*, Oxford: Windgather Press.
Bendrey, R., Hayes, T.E. and Palmer, M.R. (2009), 'Patterns of Iron Age Horse Supply: An Analysis of Strontium Isotope Ratios in Teeth', *Archaeometry*, 51(1): 140–50.
Bendrey, R., Thorpe, N., Outram, A. and van Wijngaarden-Bakker, L.H. (2013), 'The Origins of Domestic Horses in North-west Europe: New Direct Dates on the Horses of Newgrange, Ireland', *Proceedings of the Prehistoric Society*, doi:10.1017/ppr.2013.3.
Benecke, N. (1994), *Der Mensch und seine Haustiere: Die Geschichte einer jahrtausendealten Beziehung* [Man and his domestic animals: The story of a thousand-year-old relationship], Stuttgart: Theiss.
Benedictow, O.J. (2004), *The Black Death, 1346–1353: The Complete History*, Woodbridge, UK: Boydell and Brewer.
Berger, T.E., Peters, J. and Grupe, G. (2010), 'Life History of a Mule (*c*. 160 AD) from the Roman Fort *Biriciana*/Weißenburg (Upper Bavaria) as Revealed by Serial Stable Isotope Analysis of Dental Tissues', *International Journal of Osteoarchaeology*, 20: 158–71.
Binford, L.R. (1978), *Nunamiut Ethnoarchaeology*, New York: Academic Press.
Binford, L.R. (1981), *Bones: Ancient Men and Modern Myths*, New York: Academic Press.
Binois, A., Wardius, C., Rio, P., Brisault, A. and Petit, C. (2013), 'A Dog's Life: Multiple Trauma and Potential Abuse in a Medieval Dog from Guimps (Charante, France)', *International Journal of Palaeopathology*, 3: 39–47.
Birrell, J. (1982), 'Who Poached the King's deer? A Study in 13th-Century Crime', *Midland History*, 7: 9–25.
Birrell, J. (1992), 'Deer and Deer Farming in Medieval England', *Agricultural History Review*, 40(2): 112–26.
Birrell, J. (1996), 'Peasant Deer Poachers in the Medieval Forest', in R. Britnell and J. Hatcher (eds), *Progress and Problems in Medieval England: Essays in Honour of Edward Miller*, Cambridge: Cambridge University Press.
Bocheński, Z.M. and Tomek, T. (2009), *A Key for the Identification of Domestic Bird Bones in Europe: Preliminary Determination*, Kraków: Institute of Systematics and Evolution of Animals of the Polish Academy of Sciences.
Bocherens, H. and Drucker, D. (2003), 'Trophic Level Isotopic Enrichment of Carbon and Nitrogen in Bone Collagen: Case Studies from Recent and Ancient Terrestrial Ecosystems', *International Journal of Osteoarchaeology*, 13(1–2), 46–53.
Bocquet-Appel J.P. (2011a), 'When the World's Population Took Off: The Springboard of the Neolithic Demographic Transition', *Science*, 333(6042): 560–1.

Bocquet-Appel, J. P. (2011b), 'The Agricultural Demographic Transition During and After the Agriculture Inventions', *Current Anthropology*, 52(4): 497–510.

Bodson, L. (2000), 'Motivations for Pet-Keeping in Ancient Greece and Rome: A Preliminary Survey', in A.L. Podberscek, E.S. Paul and J.A. Serpell (eds), *Companion Animals and Us: Exploring the Relationship Between People and Pets*, Cambridge: Cambridge University Press.

Bogaard, A. (2012), 'Middening and Manuring in Neolithic Europe: Issues of Plausibility, Intensity and Archaeological Method', in R.L.C. Jones (ed.), *Manure Matters: Historical, Archaeological and Ethnographic Perspectives*, Farnham: Ashgate.

Bogucki, P.I. (1984), 'Ceramic Sieves of the Linear Pottery Culture and their Economic Implications', *Oxford Journal of Archaeology*, 3(1): 15–30.

Bökönyi, S. (1969), 'Archaeological Problems and Methods of Recognizing Animal Domestication', in P.J. Ucko and G.W. Dimbleby (eds), *The Domestication and Exploitation of Plants and Animals*, Chicago: Aldine Publishing.

Bökönyi, S. (1977), 'The Introduction of Sheep Breeding to Europe', *Ethnozootechnie*, 21: 65–70.

Bond, J.M. (1996), 'Burnt Offerings: Animal Bone in Anglo-Saxon Cremations', *World Archaeology*, 28(1): 76–88.

Bond, J.M. and Worley, F.L. (2006), 'Companions in Death: The Roles of Animals in Anglo-Saxon and Viking Cremation Rituals in Britain', in R. Gowland and C. Knüsel (eds), *Social Archaeology of Funerary Remains*, Oxford: Oxbow.

Bonnechere, P. (2007), 'Divination', in D. Ogden (ed.), *A Companion to Greek Religion*, Chichester, UK: Wiley-Blackwell.

Borges, J.L. (1999), *Borges: Selected Fictions*, London: Penguin.

Borić, D. (2007), 'Images of animality: Hybrid bodies and mimesis in early prehistoric art', in C. Renfrew and I. Morley (eds), *Image and Imagination: A Global Prehistory of Figurative Representation*. Cambridge: The McDonald Institute for Archaeological Research, pp. 89–105.

Boyle, J.A. (1973), 'The Hare in Myth and Reality: A Review Article', *Folklore*, 84(4): 313–26.

Bradley, R. (1978), *The Prehistoric Settlement of Britain*, London: Routledge and Kegan Paul.

Bradley, R. and Fraser, E. (2010), 'Bronze Age Barrows on the Heathlands of Southern England: Construction, Forms and Interpretations', *Oxford Journal of Archaeology*, 29(1): 15–33.

Bradley, R. and Gordon, K. (1988), 'Human Skulls from the River Thames: Their Dating and Significance', *Antiquity*, 62(236): 503–9.

Brain, C.K. (1981), *The Hunters of the Hunted? An Introduction to African Cave Taphonomy*, Chicago: University of Chicago Press.

Brandt, L.Ø., Tranekjer, L.D., Mannering, U., Ringgaard, M., Frei, K.M., Willerslev, E. and Gilbert, M.T.P. (2011), 'Characterising the Potential of Sheep Wool for Ancient DNA Analyses', *Archaeological and Anthropological Sciences*, 3(2): 209–21.

Brisbane, M., Hambleton, E., Maltby, M. and Nosov, E. (2007), 'A Monkey's Tale: The Skull of a Macaque Found at Ryurik Gorodishche During Excavations in 2003', *Medieval Archaeology*, 51: 185–90.

Brittain, M. and Overton, N. (2013), 'The Significance of Others: A Prehistory of Rhythm and Interspecies Participation', *Society and Animals*, 21: 134–49.

Britton, K., Müldner, G. and Bell, M. (2008), 'Stable Isotope Evidence for Salt-marsh Grazing in the Bronze Age Severn Estuary, UK: Implications for Palaeodietary Analysis at Coastal Sites', *Journal of Archaeological Science*, 35: 2111–18.

Britton, K.H., Grimes, V., Dau, J. and Richards, M.P. (2009), 'Reconstructing Faunal Migrations using Intra-tooth Sampling and Strontium and Oxygen Isotope Analyses: A Case Study of Modern Caribou (*Rangifer tarandus granti*)', *Journal of Archaeological Science*, 36(5): 1163–72.

Britton, K.H., Grimes, V., Niven, L., Steele, T., McPherron, S., Soressi, M., Kelly, T., Jaubert, J., Hublin, J-J. and Richards, M. (2011), 'Strontium Isotope Evidence for Migration in Late Pleistocene Rangifer: Implications for Neanderthal Hunting Strategies at the Middle Palaeolithic site of Jonzac, France', *Journal of Human Evolution*, 61(2): 176–85.

Brockman, N. (2011), *Encyclopedia of Sacred Places*, Santa Barbara, CA: ABC-CLIO.

Brown, D. (2006), 'Astral Divination in the Context of Mesopotamian Divination, Medicine, Religion, Magic, Society and Scholarship', *East Asian Science, Technology, and Medicine*, 25: 69–126.

Brownstein, O. (1969), 'The Popularity of Baiting in England Before 1600: A Study in Social and Theatrical History', *Educational Theatre Journal*, 21(3): 237–50.

Brück, J. (1999), 'Ritual and Rationality: Some Problems of Interpretation in European Archaeology', *European Journal of Archaeology*, 2(3): 313–44.

Brück, J. (2006a), 'Death, Exchange and Reproduction in the British Bronze Age', *Journal of European Archaeology*, 9(1): 73–101.

Brück, J. (2006b), 'Fragmentation, Personhood and the Social Construction of Technology in Middle and Late Bronze Age Britain', *Cambridge Archaeological Journal*, 16(3), 297–315.

Budiansky, S. (1992), *The Covenant of the Wild: Why Animals Chose Domestication*, New York: William Morrow.

Bulliet, R.W. (2005), *Hunters, Herder, and Hamburgers: The Past and Future of Human-Animal Relationships*, New York: Columbia University Press.

Buitenhuis, H. (1997), 'Aşikli Höyük: A 'Protodomestication' Site', *Anthropozoologica*, 25–26: 655–62.

Burger, J., Kirchner, M., Bramanti, B., Haak, W., and Thomas, M.G. (2007), 'Absence of the Lactase-persistence-associated Allele in Early Neolithic Europeans', *Proceedings of the National Academy of Sciences*, 104(10): 3736–41.

Burnett, C. (1997), 'An Islamic Divinatory Technique in Medieval Spain', in D.A. Agius and R. Hitchcock (eds), *The Arab Influence in Medieval Europe*, Reading, UK: Ithaca.

Burt, J. (2006), 'Conflicts Around Slaughter in Modernity', in The Animal Studies Group (eds), *Killing Animals*, Chicago: University of Illinois.

Busatta, S. (2007), 'Good To Think: Animals and Power', *Antrocom*, 4(1): 3–11.

Butler, R. (2008), *The Longevity Revolution: The Benefits and Challenges of Living a Long Life*, New York: Public Affairs.

Calver, M.C., Grayson, J., Lilith, M. and Dickman, C.R. (2011), 'Applying the Precautionary Principle to the Issue of Impacts by Pet Cats on Urban Wildlife', *Biological Conservation*, 144(6): 1895–901.

Cameron, M.L. (1993), *Anglo-Saxon Medicine*, Cambridge: Cambridge University Press.

Cannon, D.Y. (1987), *Marine Fish Osteology: A Manual for Archaeologists*, Burnaby, Canada: Simon Fraser University.

Cantor, L.M. and Wilson, J.D. (1961), 'The Mediaeval Deer-parks of Dorset: I', *Dorset Natural History and Archaeology Society*, 83: 109–16.

Carr, G. and Knüsel, C. (1997), 'The Ritual Framework of Excarnation by Exposure as the Mortuary Practice of the Early and Middle Iron Ages of Central Southern Britain', in A. Gwilt and C. Haselgrove (eds), *Reconstructing Iron Age Societies: New Approaches to the British Iron Age*, Oxford, Oxbow.

Cartmill, M. (1993), *A View to a Death in the Morning: Hunting and Nature Through History*, Cambridge: Harvard University Press.
Cauvin, J. (2000), *The Birth of the Gods and the Origins of Agriculture*, Cambridge: Cambridge University Press.
Chamberlain, N. (2005), *The Samburu in Kenya: A Changing Picture*, Unpublished report for the University of Newcastle, UK.
Chamberlain, A. (2006), *Demography in Archaeology*, Cambridge: Cambridge University Press.
Chaplin, R.E. (1971), *The Study of Animal Bones From Archaeological Sites*, London: Seminar Press.
Chapman, N. and Chapman, D. (1975), *Fallow Deer: Their History, Distribution and Biology*, Lavenham, UK: Terence Dalton.
Choyke, A. (2010), 'The bone is the beast: animal amulets and ornaments in power and magic', in D. Campana, P. Crabtree, S.D. deFrance, J. Lev-Tov and A.M. Choyke (eds), *Anthropological Approaches to Zooarchaeology: Colonialism, Complexity and Animal Transformations*, Oxford: Oxbow.
Clavel, B. (2001), 'L'Animal dans l'Alimentation Médiévale et Moderne en France du Nord (XIIe - XVIIe Siècles)', *Revue Archéologique de Picardie*, 19: 9–204.
Clavel B., Marinval-Vigne M.C., Lepetz S. and Yvinec J.-H. (1997), 'Évolution de la taille et de la morphologie du coq au cours de périodes historiques en France du Nord', *Ethnozootechnie*, 58: 3–12.
Clutton-Brock, J. (1994), 'The Unnatural World: Behavioural Aspects of Humans and Animals in the Process of Domestication', in A. Manning and J. Serpell (eds), *Animals in Human Society: Changing Perspectives*, London: Routledge.
Coates P. (1998), *Nature: Western Attitudes Since Ancient Times*, Cambridge: Polity Press.
Cobb, R. (2003), 'Chickenfighting for the Soul of the Heartland', *Text, Practice, Performance*, 4: 69–83.
Cohen, A. and Serjeantson D. (1996), *A Manual for the Identification of Bird Bones from Archaeological Sites* (2nd edn), London: Archetype.
Collins, B.J. (2002), 'Necromancy, Fertility and the Dark Earth: The Use of Ritual Pits in Hittite Cult', in P. Mirecki and M. Meyer (eds), *Magic and Ritual in the Ancient World*, Leiden: Brill.
Collins, D. (2008), 'Mapping the Entrails: The Practice of Greek Hepatoscopy', *American Journal of Philology*, 129(3): 319–45.
Conneller, C. (2004), 'Becoming Deer: Corporeal Transformation at Star Carr', *Archaeological Dialogues*, 11(1): 37–56.
Conneller, C. (2011), 'The Mesolithic', in T. Insoll (ed.), *The Oxford Handbook of the Archaeology of Ritual and Religion*, Oxford: Oxford University Press.
Conroy, D. (2004), 'Ox Yokes: Culture, Comfort and Animal Welfare', *Common Ground: Moving Forward with Animals*, On-line publication of 2004 TAWS workshop http://www.taws.org/TAWSworkshops.htm.
Cool, H.E.M. (2004), 'Some notes on spoons and mortaria', in B. Croxford, H. Eckardt, J. Meake and J. Weekes (eds), *TRAC 2003: Proceedings of the Thirteenth Annual Theoretical Roman Archaeology Conference*, Oxford: Oxbow.
Cool, H. (2006), *Eating and Drinking in Roman Britain*, Cambridge: Cambridge University Press.
Copley, M.S., Berstan, R., Dudd, S.N., Aillaud, S., Mukherjee, A.J., Straker, V., Payne, S. and Evershed, R.P. (2005), 'Processing of Milk Products in Pottery Vessels Through British Prehistory', *Antiquity*, 79(306), 895–908.

Copley, M.S., Berstan, R., Dudd, S.N., Docherty, G., Mukherjee, A.J., Straker, V. and Evershed, R.P. (2003), 'Direct Chemical Evidence for Widespread Dairying in Prehistoric Britain', *Proceedings of the National Academy of Sciences*, 100(4): 1524–9.

Cornwall, I.W. (1956), *Bones for the Archaeologist*, London: Phoenix House.

Coss, P. (2003), 'Knighthood, Heraldry and Social Exclusion in Edwardian England', in P. Coss and M. Keen (eds), *Heraldry, Pageantry and Social Display in Medieval England*, Woodbridge, UK: Boydell.

Costa-Neto, E.M. (2005), 'Animal-based Medicines: Biological Prospection and the Sustainable use of Zootherapeutic Resources', *Anais da Academia Brasileira de Ciências*, 77(1): 33–43.

Costin, C.L. (2012), 'Gender and Textile Production in Prehistory', in D. Bolger (ed.), *A Companion to Gender Prehistory*, Chichester, UK: Wiley-Blackwell.

Cotton, J., Elsden, N., Pipe, A., and Rayner, L. (2006), 'Taming the Wild: A Final Neolithic/ Earlier Bronze Age Aurochs Deposit from West London', in D. Serjeantson and D. Field (eds), *Animals in the Neolithic of Britain and Europe*, Oxford: Oxbow.

Coulton, G.G. (1925), *The Medieval Village*, Cambridge: Cambridge University Press.

Crabtree, P. (1991a), 'Zooarchaeology and Complex Societies: Some uses of Faunal Remains for the Study of Trade, Social Status and Ethnicity', *Archaeological Method and Theory*, 2: 155–205.

Crabtree, P. (1991b), 'The Symbolic Role of Animals in Anglo-Saxon England: Evidence from Burials and Cremations', in K. Ryan and P. J. Crabtree (eds), *The Symbolic Role of Animals in Archaeology*, Philadelphia: University of Pennsylvania Press.

Craig, O.E., Chapman, J., Heron, C., Willis, L. H., Bartosiewicz, L., Taylor, G., and Collins, M. (2005), 'Did the First Farmers of Central and Eastern Europe Produce Dairy Foods?', *Antiquity*, 79: 882–94.

Craig, R., Knüsel, C. J. and Carr, G.C. (2005), 'Fragmentation, Mutilation and Dismemberment: An Interpretation of Human Remains on Iron Age sites', in M. Parker Pearson and I.J. Thorpe (eds), *Warfare, Violence and Slavery in Prehistory*, (BAR International Series, 1374), Oxford: Archaeopress.

Crate, S.A. (2008), 'Walking Behind the Old Women: Sacred Sakha Cow Knowledge in the 21st Century', *Human Ecology Review*, 15(2): 115–29.

Crawford, R.D. (2003), *Poultry Breeding and Genetics*, Oxford: Elsevier.

Creighton, J. (1995), 'Visions of Power: Imagery and Symbols in Late Iron Age Britain', *Britannia*, 26: 285–301.

Creighton, J. (2000), *Coins and Power in Late Iron Age Britain*, Cambridge: Cambridge University Press.

Creighton O. (2009), *Designs Upon the Land: Elite Landscapes of the Middle Ages*, Cambridge: Boydell.

Cross, P. (2011), 'Horse Burial in First Millennium AD Britain: Issues of Interpretation', *European Journal of Archaeology*, 14(1): 190–209.

Crummy, N. (2007), 'Brooches and the Cult of Mercury', *Britannia*, *38*, 225–230.

Crummy, N. (2010), 'Bears and Coins: The Iconography of Protection in Late Roman Infant Burials', *Britannia*, 41: 37–93.

Crummy, N. (2013), 'Attitudes to the Hare in Town and Country' in Eckardt, H. and Rippon, S. (eds), *Living and Working in the Roman World: Essays in Honour of Michael Fulford on His 65th Birthday*, Dexter, MI: Thomson-Shore.

Cucchi, T., Vigne, J-D. and Auffray, J.C. (2005), 'First Occurrence of the House Mouse (*Mus Musculus Domesticus* Schwarz& Schwarz, 1943) in Western Mediterranean: A

Zooarchaeological Revision of Sub-Fossil Occurrences', *Biological Journal of the Linnean Society*, 84(3): 429–45.
Cummins, J. (1988), *The Hound and the Hawk: The Art of Medieval Hunting*, London: Weidenfield and Nicholson.
Dalley, S. (1980), 'Old Babylonian Dowries', *Iraq*, 42(1): 53–74.
Dandoy, J.R. (2006), 'Astragaloi Through Time', in M. Maltby (ed.), *Integrating Zooarchaeology*, Oxford: Oxbow.
David, B. and Thomas, J. (2008), 'Landscape Archaeology: Introduction', in B. David and J. Thomas (eds), *Handbook of Landscape Archaeology*, Walnut Creek, CA: Left Coast Press.
Davidson, H.E. (1998), *Roles of the Northern Goddess*, London: Routledge.
Davis S.J.M. (1987), *The Archaeology of Animals*, London: Batsford.
Davis, S.J.M. (1997), 'The Agricultural Revolution in England: Some Zooarchaeological Evidence', *Anthropozoologica*, 25/26: 413–28.
Davis, S.J.M. (2008a), ' "Thou Shalt Take of the Ram . . . the Right Thigh; for it is a Ram of Consecration . . .": Some Zooarchaeological Examples of Body-part Preferences', in F. D'Andria, J. De Grossi Mazzorin and G. Fiorentino (eds), *Uomini, Piante e Animali Nella Dimenione del Sacro*, Edipuglia: BACT.
Davis, S. J. (2008b), 'Zooarchaeological Evidence for Moslem and Christian Improvements of Sheep and Cattle in Portugal', *Journal of Archaeological Science*, 35(4): 991–1010.
Davis, S.J. and Beckett, J.V. (1999), 'Animal Husbandry and Agricultural Improvement: The Archaeological Evidence from Animal Bones and Teeth', *Rural History*, 10: 1–18.
Davis, S. and Payne, S. (1993), 'A Barrow Full of Cattle Skulls', *Antiquity*, 67: 12–22.
De Cupere, B., Lentacker, A., Van Neer, W., Waelkens, M., and Verslype, L. (2000), 'Osteological Evidence for the Draught Exploitation of Cattle: First Applications of a New Methodology', *International Journal of Osteoarchaeology*, 10(4): 254–67.
De Grossi Mazzorin, J. and Minniti, C. (2006), 'Dog Sacrifice in the Ancient World: A Ritual Passage?', in L.M. Snyder and E.A. Moore (eds), *Dogs and People in Social, Working Economic or Symbolic Interaction*, Oxford: Oxbow.
deFrance, SD. (2009), 'Zooarchaeology in Complex Societies: Political Economy, Status and Ideology', *Journal of Archaeological Research*, 17: 105–68.
DeJohn Anderson, V. (2004), *Creatures of Empire: How Domestic Animals Transformed Early America*, Oxford: Oxford University Press.
Dekkers, M. (2000), *Dearest Pet: On Bestiality*, London: Verso.
Derks, T. (1998), *Gods, Temples and Religious Practices: The Transformation of Religious Ideas and Values in Roman Gaul*, Amsterdam: Amsterdam University Press.
De Silva, S.S. and Turchini, G.M. (2008), 'Towards understanding the impacts of the pet food industry on world fish and seafood supplies', *Journal of Agricultural and Environmental Ethics*, 21(5), 459–67.
de Worde, W. (1508/2003), *The Boke of Keruynge* [The Book of Carving], Lewes, UK: Southover Press.
Dickinson, T.M. (2005), 'Symbols of Protection: The Significance of Animal-ornamented Shields in Early Anglo-Saxon England', *Medieval Archaeology*, 49: 109–63.
Dietler, M. and Hayden, B. (2001), *Feasts: Archaeological and Ethnographic Perspectives on Food, Politics and Power*, Washington, DC: Smithsonian Institution Press.
Dobney, K. and Ervynck, A. (2007), 'To Fish or Not to Fish? Evidence for the Possible Avoidance of Fish Consumption During the Iron Age Around the North Sea', in C. Haselgrove and T. Moore (eds), *The Later Iron Age in Britain and Beyond*, Oxford: Oxbow.

Dobney, K. and Jaques, D. (2002), 'Avian Signatures for Identity and Status in Anglo-Saxon England', *Acta Zoologica Cracoviensia*, 45: 7–21.
Doherty, S. (2013), 'New Perspectives on Cockfighting in Roman Britain', *Archaeological Review from Cambridge*, 28(2): 82–95.
Douglas, M. (1966), *Purity and Danger: An Analysis of the Concepts of Pollution and Taboo*, London: Ark Paperbacks.
Dransart, P. (2002), *Earth, Water, Fleece and Fire: An Ethnography and Archaeology of Andean Camelid Herding*, London: Routledge.
Dudd, S.N., Evershed, R.P. and Gibson, A.M. (1999), 'Evidence for Varying Patterns of Exploitation of Animal Products in Different Prehistoric Pottery Traditions Based on Lipids Preserved in Surface and Absorbed Residues', *Journal of Archaeological Science*, 26(12), 1473–82.
Dunbabin, K.M.D. (2004), *The Roman Banquet: Images of Conviviality*, Cambridge: Cambridge University Press.
Dundes, A. (1994), *The Cockfight: A Casebook*, Madison: University of Wisconsin Press.
Edwards, E. (2006), 'The Fired Clay', in E. Edwards and V. Fell (eds), *Small Finds from Eyhorne Street, Hollingbourne, CTRL Specialist Archive Report Eyhorne Street, Hollingbourne*, unpublished report for Channel Tunnel Rail Link London and Continental Railways Oxford Wessex Archaeology Joint Venture.
Elias, N. (1978), *The Civilizing Process: The History of Manners* (trans. E. Jephcott, Vol. 1), NewYork: Urizen Books. (Original work published in German in 1939).
Ellis, H. (1900), *Studies in the Psychology of Sex* (Vol. 3), New York: Random House.
Enright, M.J. (1996), *Lady with a Mead Cup: Ritual, Prophecy, and Lordship in the European Warband from La Tène to the Viking Age*, Dublin: Four Courts Press.
Entwistle, R. and Grant A. (1989), 'The Evidence for Cereal Cultivation and Animal Husbandry in the Southern British Neolithic and Bronze Age', in A. Milles (ed.), *The Beginnings of Agriculture* (British Archaeological Records International Series 496), Oxford: Archaeopress.
Erikson, P. (2000), 'The Social Significance of Pet-keeping Among Amazonian Indians', in A.L. Podberscek, E.S. Paul and J.A. Serpell (eds), *Companion Animals and Us: Exploring the Relationship Between People and Pets*, Cambridge: Cambridge University Press.
Erlandson, J.M. and Rick, T.C. (2010), 'Archaeology Meets Marine Ecology: The Antiquity of Maritime Cultures and Human Impacts on Marine Fisheries and Ecosystems', *Annual Review of Marine Science*, 2: 231–51.
Ervynck, A., Dobney, K., Hongo, H. and Meadow, R. (2001), 'Born Free? New Evidence for the Status of "*Sus scrofa*" at Neolithic Çayönü Tepesi (Southeastern Anatolia, Turkey)', *Paléorient*, 27(2): 47–73.
Eshed, V., Gopher, A., Galili, E., and Hershkovitz, I. (2004), 'Musculoskeletal Stress Markers in Natufian Hunter-gatherers and Neolithic Farmers in the Levant: The Upper Limb', *American Journal of Physical Anthropology*, 123(4): 303–15.
Evershed, R. P., Dudd, S. N., Copley, M.S. and Mukherjee, A.J. (2002), 'Identification of Animal Fats via Compound Specific $\delta^{13}C$ Values of Individual Fatty Acids: Assessments of Results for Reference Fats and Lipid Extracts of Archaeological Pottery Vessels', *Documenta praehistorica*, 21: 73–96.
Evershed, R.P., Payne, S., Sherratt, A.G., Copley, M.S., Coolidge, J., Urem-Kotsu, D., and Kostas Kotsakis, M.Ö.G. (2008), 'Earliest Date for Milk Use in the Near East and Southeastern Europe Linked to Cattle Herding', *Nature*, 455(7212): 528–31.
Everton, R.F. (1975), 'A Bos primagenuis from Charterhouse Warren Farm, Blagdon, Mendip', *Proceedings of the University of Bristol Speleological Society*, 14(1): 7–82.

Exon, S., Gaffner, V., Woodward, A. and Yorston, R. (2000), *Stonehenge Landscapes: Journeys Through Real-and-Imagined Worlds*, Oxford: Archeopress.
Fairnell, E. (2003), *The Utilisation of Fur-bearing Animals in the British Isles: A Zooarchaeological Hunt for Data*, Unpublished MSc dissertation, University of York.
Fairnell, E.H. (2008), '101 Ways to Skin a Fur-bearing Animal: The Implications for Zooarchaeological Interpretation', in P. Cunningham, J. Heeb and R. Paardekooper (eds), *Experiencing Archaeology by Experiment*, Oxford: Oxbow.
Faith, R. (1997), *The English Peasantry and the Growth of Lordship*, London: Leicester University Press.
Fang, M., Larson, G., Soares Ribeiro, H., Li, N. and Andersson, L. (2009), 'Contrasting Mode of Evolution at a Coat Color Locus in Wild and Domestic Pigs', *PLoS Genetics*, 5(1): e1000341.
Fern, C. (2010), 'Horses in Mind', in M. Carver, A. Sanmark and S. Semple (eds), *Signals of Belief in Early England: Anglo-Saxon Paganism Revisited*, Oxford: Oxbow.
Fessler, D.M. and Navarrete, C.D. (2003), 'Meat is Good to Taboo', *Journal of Cognition and Culture*, 3(1): 1–40.
Fiddes, N. (1991), *Meat*, London: Routledge.
Finet, A. (1982), 'L'Oeuf d'autruche', in J. Quaegebeur (ed.), *Studia Paulo Nasteri*, Leuven: Peeters.
Fitzhugh, W.W. (2009), 'Stone Shamans and Flying Deer of Northern Mongolia: Deer Goddess of Siberia or Chimera of the Steppe?', *Arctic Anthropology*, 46(1–2): 72–88.
Fitzpatrick, S.M. and Callaghan, R. (2009), 'Examining Dispersal Mechanisms for the Translocation of Chicken (*Gallus gallus*) from Polynesia to South America', *Journal of Archaeological Science*, 36(2): 214–23.
Fleming, A. (2010), 'Horses, Elites . . . and Long-distance Roads', *Landscapes*, 11(2): 1–20.
Fletcher, J. (2011), *Gardens of Earthly Delight: The History of Deer Parks*, Oxford: Windgather.
Forester, E.S. and Heffner, E.H. (1955), *Columella, L.J.M: On Agriculture* (Loeb Classical Library 361), London: Heinemann.
Foucault, M. (1970) *The Order of Things: An Archaeology of the Human Sciences* (trans. A. Sheridan), New York: Random House. (Original published in 1966 in French.)
Fowler, C. (2011), 'Personhood and the Body', in T. Insoll (ed.), *The Oxford Handbook of the Archaeology of Ritual and Religion*, Oxford: Oxford University Press.
Frazer, J.G. (1912), *Spirits of the Corn and of the Wild*, London: Macmillan.
Fredericks, A.D. (2010), *How Long Things Live*, Mechanicsburg, PA: Stackpole Books.
Gade, K.E. (1986), 'Homosexuality and Rape of Males in Old Norse Law and Literature', *Scandinavian Studies*, 8: 114–32.
Game, A. (2001), 'Riding: Embodying the Centaur', *Body and Society*, 7(4): 1–12.
Gardeisen, A. (2002), *Mouvements ou Déplacements de Populations Animales en Méditerranée au cours de l'Holocène* (British Archaeological Reports, International Series 1017), Oxford: Archeopress.
Gardiner, M. (1997), 'The Exploitation of Sea-mammals in Medieval England: Bones and Their Social Context', *Archaeological Journal*, 154: 173–95.
Garmonsway, G.N. (1939), *Aelfric's Colloquy*, London: Methuen.
Gautier, A. (2006), *Le festin dans l'Angleterre anglo-saxonne*, Rennes, France: Presses universitaires de Rennes.
Gautier, A. (2009), 'Hospitality in Pre-Viking Anglo-Saxon England', *Early Medieval Europe*, 17(1): 23–44.

Gazin-Schwartz, A. (2001), 'Archaeology and Folklore of Material Culture, Ritual, and Everyday Life', *International Journal of Historical Archaeology*, 5(4): 263–80.

Geertz, C. (1994), 'Deep Play: Notes on the Balinese Cockfight', in A. Dundes (ed.), *The Cockfight: A Casebook*, Madison: University of Wisconsin Press.

Gelling, M. (1987), 'Anglo-Saxon Eagles', *Leeds Studies in English*, 18: 173–81.

Genovesi, P., Bacher, S., Kobelt, M., Pascal, M. and Scalera, R. (2009), 'Alien Mammals of Europe', in J.A. Drake (ed.), *DAISIE, Handbook of Alien Species in Europe*, London: Springer.

Gilbert, J. (1979), *Hunting and Hunting Reserves in Medieval Scotland*, Edinburgh: John Donald.

Gilchrist, R. (1999), *Gender and Archaeology: Contesting the Past*, London: Routledge.

Gilchrist, R. (2008), 'Magic for the Dead? The Archaeology of Magic in Later Medieval Burials', *Medieval Archaeology*, 52(1): 119–59.

Gilchrist, R. (2012), *Medieval Life: Archaeology and the Life Course*, Woodbridge, UK: Boydell.

Gilhus, I.S. (2006), *Animals, Gods and Humans: Changing Attitudes to Animals in Greek, Roman and Early Christian Ideas*, London: Routledge.

Gilmour, G.H. (1997), 'The Nature and Function of Astragalus Bones from Archaeological Contexts in the Levant and Eastern Mediterranean', *Oxford Journal of Archaeology*, 16(2): 167–75.

Gleba, M. (2009), 'Textile Tools in Ancient Italian Votive Contexts: Evidence of Dedication or Production?', in M. Gleba and H. Becker (eds), *Votives, Places and Rituals in Etruscan Religion: Studies in Honor of Jean MacIntosh Turfa*, Leiden: Brill.

Gleba, M. (2012), 'From Textiles to Sheep: Investigating Wool Fibre Development in pre-Roman Italy using Scanning Electron Microscopy (SEM)', *Journal of Archaeological Science*, 39(12): 3643–61.

Godden, M.R. (1990), 'Money, Power and Morality in Late Anglo-Saxon England', *Anglo-Saxon England*, 19(1): 41–65.

Goering J. and Mantello, F.A.C. (1986), 'The "Perambulauit Iudas . . ." (Speculum Confessionis) Attributed to Robert Grosseteste', *Revue Benedictine*, 96: 125–68.

Good, I. (2012), 'Textiles', in D.T. Potts (ed.), *A Companion to the Archaeology of the Ancient Near East*, Chichester: Wiley-Blackwell.

Goody, J. (1982), *Cooking, Cuisine and Class: A Study in Comparative Sociology*, Cambridge: Cambridge University Press.

Gongora, J., Rawlence, N.J., Mobegi, V.A., Jianlin, H., Alcalde, J.A., Matus, J.T., Hanotte, O., Moran, C., Austin, J.J., Ulm, S., Anderson, A.J., Larson, G. and Cooper, A. (2008), Indo-European and Asian origins for Chilean and Pacific Chickens revealed by mtDNA', *PNAS*, 105: 10308–13.

Graham, J. and Haidt, J. (2010), 'Beyond Beliefs: Religions Bind Individuals into Moral Communities', *Personality and Social Psychology Review*, 14(1): 140–50.

Grant, A. (1981), 'The Significance of Deer Remains at Occupation Sites of the Iron Age to the Anglo-Saxon Period', in M. Jones and G. Dimbleby (eds), *The Environment of Man: The Iron Age to the Anglo-Saxon period* (British Archaeologocal Reports, British Series 87), Oxford: Archeopress.

Grant, A. (1982), 'The Use of Tooth Wear as a Guide to the Age of Domestic Ungulates', in B. Wilson, C. Grigson, and S. Payne (eds), *Aging and Sexing Animal Bones from Archaeological Sites*, (British Archaeological Reports, British Series 109), Oxford: Archeopress.

Grant, A. (1984), 'Survival or Sacrifice? A Critical Appraisal of Animal Burials in Britain in the Iron Age', in J. Clutton-Brock and C. Grigson (eds), *Animals and Archaeology* (British Archaeological Reports, International Series 202), Oxford: Archaeopress.

Grant, A. (1987), 'Some Observations on Butchery in England from the Iron Age to the Medieval Period', *Le découpe e le partage du corps à travers le temps et l'espace, Anthropozoologica*, 1: 53–7.

Grant, M. (2000), *Galen on Food and Diet*, London: Routledge.

Gräslund, A.S. (2004), 'Dogs in Graves: A Question of Symbolism?', in B. Santillo Frizell (ed.), *Pecus: Man and Animal in Antiquity*, Rome: the Swedish Institute of Rome.

Grayson, D.K. (2001), 'The Archaeological Record of Human Impact on Animal Populations', *Journal of World Prehistory* 15(1): 1–68.

Greaves, A.M. (2012), 'Divination at Archaic Branchigai-Didyma: A Critical Review', *Hesperia*, 81(2): 177–206.

Green, J.B. (1997), *The Gospel of Luke*, Cambridge: Eerdmans Publishing.

Green, M. (1992), *Animals in Celtic Life and Myth*, London: Routledge.

Green, N. (2006), 'Eggs and Peacock Feathers: Sacred Objects as Cultural Exchange between Christianity and Islam', *Al-Masāq*, 18(1): 27–78.

Greenfield, H.J. (1988), 'The Origin of Milk and Wool Production in the Old World', *Current Anthropology*, 29(4): 573–92.

Greenfield, H.J. (2010), 'The Secondary Products Revolution: The Past, the Present and the Future', *World Archaeology*, 42(1): 29–54.

Griffin, E. (2007), *Blood Sport: Hunting in Britain since 1066*, Bury St Edmunds, UK: Yale University Press.

Grigson, C. (1969), 'The Uses and Limitations of Differences in Absolute Size in the Distinction between the Bones of Aurochs (*Bos Primigenius*) and Domestic Cattle (*Bos Taurus*)', in P.J. Ucko and G.W. Dimbleby (eds), *The Domestication and Exploitation of Plants and Animals*, London: Duckworth.

Grigson, C. (1999), 'The Mammalian Remains', in A. Whittle, J. Pollard and C. Grigson (eds), *The Harmony of Symbols: The Windmill Hill Causeway Enclosure Wiltshire*, Oxford: Oxbow.

Groot, M. (2008), *Animals in Ritual and Economy in a Roman Frontier Community: Excavations in Tiel-Passewaaij (Vol. 12)*, Amsterdam: Amsterdam University Press.

Guggenheim, S. (1994), 'Cock or Bull: Cockfighting, Social Structure, and Political Commentary in the Philippines', in A. Dundes (ed.), *The Cockfight: A Casebook*, Madison: University of Wisconsin Press.

Guiry, E.J. (2012), 'Dogs as Analogs in Stable Isotope-based Human Palaeodietary Reconstructions: A Review and Considerations for Future Use', *Journal of Archaeological Method and Theory*, 19: 351–76.

Guiry, E.J. (2013), 'A Canine Surrogacy Approach to Human Paleodietary Bone Chemistry: Past Development and Future Directions', *Archaeological and Anthropological Sciences*, 19(3): 1–12.

Gupta, A.K. (2004), 'Origin of Agriculture and Domestication of Plants and Animals Linked to Early Holocene Climate Amelioration', *Current Science*, 87(1): 54–9.

Guy, R.D., Reid, D.M. and Krouse, H.R. (1986a), 'Factors affecting C13/C12 Ratios of Inland Halophytes. 1: Controlled Studies on Growth and Isotopic Composition of *Puccinellia Nuttalliana*', *Canadian Journal of Botany*, 64: 2693–9.

Guy, R.D., Reid, D.M. and Krouse, H.R. (1986b), 'Factors affecting C13/C12 Ratios of Inland Halophytes. 2: Ecophysiological Interpretations of Patterns in the Weld', *Canadian Journal of Botany*, 64: 2700–7.

Halstead, P. (1996), 'The Development of Agriculture and Pastoralism in Greece: When, How, Who and What', in D.R. Harris (ed.), *The Origins and Spread of Agriculture and Pastoralism in Eurasia*, London: UCL press.

Halstead, P. and Barrett, J.C. (2004), 'Introduction: Food, Cuisine and Society in Prehistoric Greece', in P. Halstead and J.C. Barrett (eds), *Food, Cuisine and Society in Prehistoric Greece*, Oxford: Oxbow, pp. 1–15.
Hambleton, E. (1999), *Animal Husbandry Regimes in Iron Age Britain* (BAR, British Series 282), Oxford: Archeopress.
Hambleton, E. (2008), *Review of Middle Bronze Age to Late Iron Age Faunal Assemblages from Southern Britain*, English Heritage Research Report 71/2008.
Hambrecht, G. (2006), 'The Bishop's Beef: Improved Cattle at Early Modern Skálholt, Iceland', *Archaeologia Islandica*, 5: 82–94.
Hambrecht, G. (2009), 'Zooarchaeology and the Archaeology of Early Modern Iceland', *Journal of the North Atlantic*, 2: 3–22.
Hamerow, H. (2006), 'Special Deposits in Anglo-Saxon Settlements', *Medieval Archaeology*, 50: 1–30.
Hamilakis, Y. (1998), 'Eating the Dead: Mortuary Feasting and the Politics of Memory in the Aegean Bronze Age Societies', in K. Branigan (ed.), *Cemetery and Society in the Aegean Bronze Age*, Sheffield: Sheffield Academic Press.
Hamilakis, Y. (2000), 'The Anthropology of Food and Drink: Consumption and Aegean Archaeology', in S.J. Vaughan and W.D. Coulson (eds), *Palaeodiet in the Aegean: Papers from a Colloquium Held at the 1993 Meeting of the Archaeological Institute of America in Washington*, Oxford: Oxbow.
Hamilakis, Y. (2003), 'The Sacred Geography of Hunting: Wild Animals, Social Power and Gender in Early Farming Societies', in E. Kotjabopoulou, Y. Hamilakis, P. Halstead, C. Gamble and V. Elafanti (eds), *Zooarchaeology in Greece: Recent Advances*, London: British School at Athens.
Hamilakis, Y. and Konsolaki, E. (2004), 'Pigs for the Gods: Burnt Animal Sacrifices as Embodiment Rituals at a Mycenaean Sanctuary', *Oxford Journal of Archaeology*, 23(2): 135–51.
Hammon, A. (2010), 'The Brown Bear', in T. O'Connor and N. Sykes (eds), *Extinctions and Invasions: A Social History of British Fauna*, Oxford: Windgather.
Handler, J.S. (1997), 'An African-type Healer/diviner and His Grave Goods: A Burial from a Plantation Slave Cemetery in Barbados, West Indies', *International Journal of Historical Archaeology*, 1(2): 91–130.
Handford S.A. (1982), *Caesar: The Conquest of Gaul*, London: Penguin.
Harcourt, R.A. (1974), 'The Dog in Prehistoric and Early Historic Britain', *Journal of Archaeological Science*, 1(2): 151–75.
Harcourt, R. (1979), 'The Animal Bones', in G.J. Wainwright and H.C. Bowen (eds), *Gussage All Saints: An Iron Age Settlement in Dorset*, London: Department of Environment Report, 10.
Harding, A.F. (2000), *European Societies in the Bronze Age*, Cambridge: Cambridge University Press.
Härke, H. (1989), 'Knives in Early Saxon Burials: Blade Lengths and Age at Death', *Medieval Archaeology*, 33: 144–8.
Härke, H. (2000) 'The Circulation of Weapons in Anglo-Saxon Society', in F. Theuws and J.L. Nelson (eds), *Rituals of Power: From Late Antiquity to the Early Middle Ages*, Leiden: Brill.
Harris M. (1974), *Cows, Pigs, Wars, and Witches: The Riddles of Culture*, New York: Random House.
Hawkes, J. (1997), 'Symbolic Lives: The Visual Evidence', in J. Hines (ed.), *The Anglo-Saxons from the Migration Period to the Eighth Century: An Ethnographic Perspective*, Woodbridge, UK: Boydell.

Hedges, R.E. and Reynard, L.M. (2007), 'Nitrogen Isotopes and the Trophic Level of Humans in Archaeology', *Journal of Archaeological Science*, 34(8): 1240–51.

Hedges, R., Saville, A. and O'Connell, T. (2008), 'Characterizing the Diet of Individuals at the Neolithic Chambered Tomb of Hazleton North, Gloucestershire, England, using Stable Isotope Analysis', *Archaeometry*, 50(1): 114–28.

Helms, M. (1993), *Craft and the Kingly Ideal: Art Trade and Power*. Austin, TX: University of Texas Press.

Helmer, D., Gourichon, L., Monchot, H., Peters, J. and Sana Segui, M. (2005) 'Identifying Early Domestic Cattle from Pre-Pottery Neolithic Sites on the Midddle Euphrates using Sexual Dimorphism', in J-D. Vigne, J. Peters and D. Helmer (eds), *The First Steps of Animal Domestication: New Archaeozoological Approaches*, Oxford: Oxbow.

Henderson, J. (2007), *The Atlantic Iron Age: Settlement and Identity in the First Milennium BC*, Oxford: Routledge.

Henrich, J. and Henrich, N. (2010), 'The Evolution of Cultural Adaptations: Fijian Food Taboos Protect Against Dangerous Marine Toxins', *Proceedings of the Royal Society B: Biological Sciences*, 277(1701): 3715–24.

Henry, B.C. (2004), 'The Relationship Between Animal Cruelty, Delinquency, and Attitudes Toward the Treatment of Animals', *Society and Animals*, 12(3): 184–207.

Henton, S., Meier-Augenstein, W. and Kemp, H.F. (2010), 'The Use of Oxygen Isotopes in Sheep Molars to Investigate Past Herding Practices at the Neolithic Settlement of Çatalhöyük, Central Anatolia', *Archaeometry*, 52(3): 429–49.

Herold, M. (2008), *Sex Differences in Mortality in Lower Austria and Vienna in the Early Medieval Period: An Investigation and Evaluation of Possible Contributing Factors*, Unpublished PhD thesis, University of Vienna.

Hernandez, J. (2010), 'A New Renaissance Medical Controversy: Sixteenth-century Polemics about Cold-drinking', in D. Collard, J. Morris and E. Perego (eds), *Food and Drink in Archaeology 3*. Totnes, UK: Prospect Books.

Herring, P. (2003), 'Cornish Medieval Deer Parks', in R. Wilson-North (ed.), *The Lie of the Land: Aspects of the Archaeology and History of the Designed Landscape in the South West of England*, Exeter, UK: The Mint Press.

Hesse, B. (1990), 'Pig Lovers and Pig Haters: Patterns of Palestinian Pork Production', *Journal of Ethnobiology*, 10(2): 196–225.

Hesse, B. and Wapnish, P. (1997), 'Can Pig Remains be used for Ethnic Diagnosis in the Ancient Near East', in N. Asher Silberman and D. Small (eds), *The Archaeology of Israel: Constructing the Past, Interpreting the Present*, Sheffield, UK: Sheffield Academic Press.

Hetherington, D.A., Lord, T.C. and Jacobi, R.M. (2006), 'New Evidence for the Occurrence of Eurasian Lynx (*Lynx lynx*) in Medieval Britain', *Journal of Quaternary Science*, 21(1): 3–8.

Hiatt, A. (2000), 'The Cartographic Imagination of Thomas Elmham', *Speculum*, 75(4): 859–86.

Hicks, D. (2006/7), 'Blood, Violence and Gender Alignment: Cockfighting and Kickfighting in East Timor', *Cambridge Anthropology*, 26(3): 1–20.

Hill, J.D. (1995), *Ritual and Rubbish in the Iron Age of Wessex: A Study on the Formation of a Specific Archaeological Record* (British Archaeological Reports, British Series 242), Oxford: Archeopress.

Hillson, S. (1992), *Mammal Bones and Teeth: An Introductory Guide to Methods of Identification*, London: Institute of Archaeology.

Hingley, R. (2006), 'The Deposition of Iron Objects in Britain During the Later Prehistoric and Roman Periods', *Britannia*, 37: 213–57.

Hinton, D. A. (2005), *Gold and Gilt, Pots and Pins: Possessions and People in Medieval Britain*, Oxford: Oxford University Press.

Hitchcock, L.A., Laffineur R. and Crowley, J. (2008), *DAIS. The Aegean Feast*, Liège and Austin: Aegaeum 29.

Hodder, I. (1990), *The Domestication of Europe: Structure and Contingency in Neolithic Societies*, Oxford: Basil Blackwell.

Hodkinson, P. (2013), *Feeling Cocky? A Gendered Study of Human-chicken Relationships in Medieval England*. Unpublished BA dissertation, University of Nottingham.

Holden, C.J. and Mace, R. (2003), 'Spread of Cattle led to the Loss of Matrilineal Descent in Africa: A Coevolutionary Analysis', *Proceedings of the Royal Society of London. Series B: Biological Sciences*, 270(1532): 2425–33.

Hollimon, S.E. (2001), 'The Gendered Peopling of North America: Addressing the Antiquity of Systems of Multiple Genders', in N. Price (ed.), *The Archaeology of Shamanism*, London: Routledge.

Holloway, R.R. (1994), *The Archaeology of Early Rome and Latium*, London: Routledge.

Hooke, D. (1989), 'Pre-conquest Woodland: Its Distribution and Usage', *Agricultural History Review*, 37(2): 113–29.

Hooke, D. (1998), *The Landscape of Anglo-Saxon England*, London: Leicester University Press.

Hooper, W.D. and Ash, H.B. (1979) *Marcus Terentius Varro On Agriculture* (Loeb Classical Library 283), London: Harvard University Press.

Horwitz, L.K. and Goring-Morris, N. (2004), 'Animals and Ritual during the Levantine PPNB: A Case Study from the Site of Kfar Hahoresh, Israel', *Anthropozoologica*, 39(1): 165–78.

Howard, W. (2013), *Commensal or Comestible? The Role and Exploitation of Small, Non-ungulate Mammals in Early European Prehistory: Towards a Methodology for Improving Identification of Human Utilisation*, Unpublished PhD thesis, University of Exeter, UK.

Howcroft, R. (2012), 'From Breast to Beast?: Exploring the Relationship between Dairy Farming, Infant Feeding and Demographic Change', in LeCHE (eds), *May Contain Traces of Milk: Investigating the Role of Dairy Farming and Milk Consumption in the European Neolithic*, Heslington: University of York.

Howey M.C.L. and O'Shea, M. (2006), 'Bear's Journey and the Study of Ritual in Archaeology', *American Antiquity*, 71: 261–82.

Hu, Y., Hu, S., Wang, W., Wu, X., Marshall, F. B., Chen, X. and Wang, C. (2014), 'Earliest Evidence for Commensal Processes of Cat Domestication', *Proceedings of the National Academy of Sciences*, 111(1): 116–20.

Huff, D. (2004), 'Archaeological Evidence for Zoroastrian Funerary Practices', in M. Stassberg (ed.), *Zoroastrian Rituals in Context*, Leiden: Brill.

Hughes, J.D. (2003), 'Europe as Consumer of Exotic Biodiversity: Greek and Roman Times', *Landscape Research*, 28(1): 21–31.

Hulme, P.E. (2007), 'Biological Invasions in Europe: Drivers, Pressures, States, Impacts and Responses', in R.E. Hester and R.M. Harrison (eds), *Biodiversity Under Threat*, Cambridge: Royal Society of Chemistry.

Ingold, T. (1994), 'From Trust to Domination: An Alternative History of Human-animal Relations', in A. Manning and J. Serpell (eds), *Animals and Human Society: Changing Perspectives*, London: Routledge.

Ingold T. (2000), *The Perception of the Environment: Essays in Livelihood, Dwelling and Skill*, London: Routledge.

Ingram, C.J., Raga, T.O., Tarekegn, A., Browning, S.L., Elamin, M.F., Bekele, E. and Swallow, D.M. (2009), 'Multiple Rare Variants as a Cause of a Common Phenotype: Several Different Lactase Persistence Associated Alleles in a Single Ethnic Group', *Journal of molecular evolution*, 69(6): 579–88.
Insoll, T. (2004), 'Syncretism, Time and Identity: Islamic Archaeology in West Africa', in D. Whitcomb (ed.), *Changing Social Identity with the Spread of Islam, Archaeological Perspectives*, Chicago: the Oriental Institute.
Insoll, T. (2011), *The Oxford Handbook of the Archaeology of Ritual and Religion*, Oxford: Oxford University Press.
Ijzereef, G.F. (1989), 'Social Differentiation from Animal Bone Studies', in D. Serjeantson and T. Waldron (eds), *Diet and Crafts in Towns: The Evidence of Animal Remains the Roman to the Post-Medieval Periods* (British Archaeological Reports, British Series 199), Oxford: Archeopress.
Isaakidou, V. (2006), 'Ploughing with Cows: Knossos and the Secondary Products Revolution' in D. Serjeantson and D. Field (eds), *Animals in the Neolithic of Britain and Europe*, Oxford: Oxbow.
Jacobi, K.P. (2003), 'Body Disposition in Cross-cultural Context', in Bryant C.D. (ed.), *Handbook of Death and Dying* (Vol. 1), London: Sage.
Jay, M. and Richards, M.P. (2006), 'Diet in the Iron Age Cemetery Population at Wetwang Slack, East Yorkshire, UK: Carbon and Nitrogen Stable Isotope Evidence', *Journal of Archaeological Science*, 33: 653–62.
Jennbert, K. (2011), *Animals and Humans: Recurrent Symbiosis in Archaeology and Old Norse Religion*, Lund: Nordic Academic Press.
Jesch, J. (1991), *Women in the Viking Age*. Woodbridge, UK: Boydell.
Johannsen, N.N. (2006), 'Draught Cattle and the South Scandinavian Economies of the 4th Millennium BC', *Environmental Archaeology*, 11(1): 35–48.
Johnson, M. (2007), *Ideas of Landscape*, London: Blackwell.
Johnson, M.P. (2006), 'Conflict and Control: Gender Symmetry and Asymmetry in Domestic Violence', *Violence Against Women*, 12(11): 1003–18.
Johnston, S.I. (2009), *Ancient Greek Divination*, Oxford: Blackwell.
Jolly, K., Raudvere, C. and Peters, E. (2002), *Witchcraft and Magic in Europe, Volume 3: The Middle Ages*, London: Athlone Press.
Jones, A. (2002), 'A Biography of Colour: Colour, Material Histories and Personhood in the Early Bronze Age of Britain and Ireland', in A. Jones and G. MacGregor (eds), *Colouring the Past: The Significance of Colour in Archaeological Research*, Oxford: Berg.
Jones, C.P. (1987), 'Stigma: Tattooing and Branding in Graeco-Roman Antiquity', *The Journal of Roman Studies*, 77: 139–55.
Jones, R. (2013), *The Medieval Natural World*, Harlow: Pearson.
Jones, S. (1997), *The Archaeology of Ethnicity: Constructing Identities in the Past and Present*, London: Routledge.
Jones, W.H.S. (1963), *Pliny Natural History, Volume VIII, Books XXVIII–XXXII*, London: Harvard University Press.
Jørgensen, L.B. (2003), 'Europe', in D.T. Jenkins (ed.), *The Cambridge History of Western Textiles* (Vol. 1), Cambridge: Cambridge University Press.
Joy, M. (2009), *Why We Love Dogs, Eat Pigs, and Wear Cows: An Introduction to Carnism*, San Francisco: Conari Press.
Kalof, L. (2007), *Looking at Animals in Human History*, London: Reaktion Books.
Keats-Rohan, K.S.B. (1993), *John of Salisbury, Policraticus I–IV*, Turnhout, Belgium: Brepols.

Kehoe, A.B. (2000), 'Mississippian Weavers', in A.E Rautman (ed.), *Reading the Body: Representations and Remains in the Archaeological Record*, Philadelphia: University of Pennsylvania Press.

Keith, K. (1998), 'Spindle Whorls, Gender, and Ethnicity at Late Chalcolithic Hacinebi Tepe', *Journal of Field Archaeology*, 25(4): 497–515.

Kerridge, E. (1967), *The Agricultural Revolution*, London: Allen and Unwin.

Kessinger, K.M. (1989), 'Hunting and Male Domination in Cashinahua Society', in S. Kent (ed.), *Farmers as Hunters: The Implications of Sedentism*, Cambridge: Cambridge University Press.

King, A. (2005), 'Animal Remains from Temples in Roman Britain', *Britannia*, 36: 329–69.

Kitchener, A. (2010), 'The Elk', in T. O'Connor and N. Sykes (eds), *Extinctions and Invasions: A Social History of British Fauna*, Oxford: Windgather.

Kline, N.R. (2001), *Maps of Medieval Thought*, Woodbridge, UK: Boydell.

Klingender, F. (1971), *Animals in Art and Thought to the End of the Middle Ages*, London: Routledge.

Knight, J. (2005), *Animals in Person: Cultural Perspectives on Human-Animal Intimacies*, Oxford: Berg.

Koehl, R.B. (1986), 'The Chieftain Cup and a Minoan Rite of Passage', *The Journal of Hellenic Studies*, 106: 99–110.

Koerper, H.C. and Whitney-Desautels, N. (1999), 'Astragalus Bones: Artifacts or Ecofacts', *Pacific Coast Archaeological Society Quarterly*, 35(2–3): 69–80.

Köhler-Rollefson, I. and Rollefson, G.O. (2002), 'Brooding About Breeding: Social Implications for the Process of Animal Domestication', in R.T.J. Cappers and S. Bottema (eds), *The Dawn of Farming in the Near East*, Berlin: ex Oriente.

Kothari, B. and Sharma, B.K. (2013), 'An Anthropological Account of Bonhomie and Opprobrium Between Communities and Animals in Rajasthan', in B.K. Sharma, S. Kulshreshtha and A.R. Rahmani (eds), *Faunal Heritage of Rajasthan, India: Ecology and Conservation of Vertebrates*, New York: Springer.

Kristoffersen, S. (2010), 'Half Beast-Half Man: Hybrid Figures in Animal Art', *World Archaeology*, 42(2): 261–72.

Kroll, H. (2013), 'Ihrer Hühner waren drei und ein stolzer Hahn dabei – Überlegungen zur Beigabe von Hühnern im awarischen Gräberfeld an der Wiener Csokorgassemore', *Offa*, 69/70: 201–16.

Kulakov, V.I. and Markovets, M.Y. (2004), 'Birds as Companions of Germanic Gods and Heroes', *Acta Archaeologica*, 75(2): 179–88.

Kyle, D.G. (1998), *Spectacles of Death in Ancient Rome*, London: Routledge.

Kyle, D.G. (2003), 'From the Battlefield to the Arena: Gladiators, Militarism and the Roman Republic', in J.A. Mangan (ed.), *Militarism, Sport, Europe: War Without Weapons*, London: Frank Cass Publishers.

Lage, W. (2009), 'Schleifknochen versus Schabbahnknochen: Untersuchungen zur Verwendung steinzeitlicher Langknochen von Großsäugern mit konkaven Arbeitsbahnen', *Schriften des Naturwissenschaftlichen Vereins für Schleswig-Holstein*, 71: 26–40.

Lambert R.A and Rotherham I.D. (2011), *Invasive and Introduced Plants and Animals: Human Perceptions, Attitudes and Approaches to Management*, London: Earthscan.

Landais, E. (2001), 'The Marking of Livestock in Traditional Pastoral Societies', *Revue Scientifique et Technique (International Office of Epizootics)*, 20(2): 463–79.

Larrère, C. and Larrère, R. (2000), 'Animal Rearing as a Contract?', *Journal of Agricultural and Environmental Ethics*, 12(1): 51–8.

Larson, G. (2011), 'Genetics and Domestication Important Questions for New Answers', *Current Anthropology*, 52: 485–95.
Larson, G. and Burger, J. (2013), 'A Population Genetics View of Animal Domestication', *Trends in Genetics*, 29(4), 197–205.
Larson, G., Karlsson, E., Perri, A., Webster, M., Ho, S.Y.W., Peters, J., Stahl, P.W., Piper, P.J., Lingaas, F., Fredholm, M., Comstock, K.E., Modiano, J.F., Schelling, C., Agoulnik, A.I., Leegwater, P., Dobney, K., Vigne, J-D., Vilà, C., Andersson, L. and Lindblad-Toh, K. (2012), 'Dog Domestication Revisited: A New Genetic, Archeological, and Biogeographic Perspective', *Proceedings of the National Academy of Sciences*, 109(23): 8878–83.
Larson, J. (in press), 'Venison for Artemis? The Problem of Deer Sacrifice', in S. Hitch and I. Rutherford (eds), *Animal Sacrifice in Ancient Greece*, Cambridge: Cambridge University Press.
Laurioux, B. (1988), 'Le Lièvre Lubrique et la Bête Sanglante: Réflexions sur quelques Interdits Alimentaires du Haut Moyen Âge', *Anthropozoologica*, 2: 127–32.
Lauwerier R.C.G.M. and Plug, I. (2005), *The Future from the Past: Archaeozoology in Wildlife Conservation and Heritage Management*, Oxford: Oxbow.
Leach, E.E. (2006), '"The Little Pipe Sings Sweetly while the Fowler Deceives the Bird": Sirens in the Later Middle Ages', *Music and Letters*, 87(2): 187–211.
Leach, H.M. (2003), 'Human Domestication Reconsidered', *Current Anthropology*, 44(3): 349–68.
Leduc, C. (2012), 'New Mesolithic Hunting Evidence from Bone Injuries at Danish Maglemosian Sites: Lundby Mose and Mullerup (Sjælland)', *International Journal of Osteoarchaeology*, doi: 10.1002/oa.2234.
Lee, C. (2007), *Feasting the Dead: Food and Drink in Anglo-Saxon Burial Rituals*, Woodbridge, UK: Boydell.
Legge, A.J. (1992), *Animals, Environment and the Bronze Age Economy. Fascicule 4 of Excavations at Grimes Graves, Norfolk 1972–1976*, London: The British Museum.
Legge, A.J. (2006), 'Milk use in Prehistory: the osteological evidence', in J. Mulville and A.K. Outram (eds), *The Zooarchaeology of Fats, Oils, Milk and Dairying*, Oxford: Oxbow.
Legge, A.J. (2010), 'The Aurochs', in T. O'Connor and N. Sykes (eds), *Extinctions and Invasions: A Social History of British Fauna*, Oxford: Windgather.
Legge, A.J. and Rowley-Conwy, P. (1988), *Star Carr Revisited: A Re-analysis of the Large Mammals*, London: Birkbeck College.
Leonardi, M., Gerbault, P. Thomas, M.G. and Burger, J. (2012), 'The Evolution of Lactase Persistence in Europe: A Synthesis of Archaeological and Genetic Evidence', *International Dairy Journal*, 22: 88–97.
Lev, E. (2003), 'Traditional Healing with Animals (Zootherapy): Medieval to Present-day Levantine Practice', *Journal of Ethnopharmacology*, 85(1): 107–18.
Lever, C. (2009), *The Naturalised Animals of the British Isles*. St Albans, UK: Granada Publishing.
Lévi-Strauss, C. (1969), *The Raw and the Cooked: Introduction to a Science of Mythology*, New York: Harper and Row.
Lewis, M. (2007), 'Identity and Status in the Bayeux Tapestry: The Iconographic and Artefactual Evidence', *Anglo-Norman Studies*, 29: 100–20.
Liddiard, R. (2000), *Landscapes of Lordship: The Castle and the Countryside in Medieval Norfolk, 1066–1200* (British Archaeological Reports British Series 309), Oxford: Archeopress.
Liddiard, R. (2003), 'The Deer Parks of Domesday Book', *Landscape*, 4(1), 4–23.

Liddiard, R. (2005), *Castles in Context: Power, Symbolism and Landscape, 1066 to 1500*, Macclesfield, UK: Windgather Press.

Lightfoot, E., O'Connell, T.C., Stevens, R.E., Hamilton, J., Hey, G. and Hedges, R.E.M. (2009), 'An Investigation into Diet at the Site of Yarnton, Oxfordshire, using Stable Carbon and Nitrogen Isotopes', *Oxford Journal of Archaeology*, 28(3): 301–22.

Ling, J. (1997), *A History of European Folk Music*, Rochester, NY: University Rochester Press.

Liu, Y-P., Wu, G-S., Yao, Y-G., Miao, Y-W., Luikart, G., Baig M., Beja-Pereira, A., Ding, Z-L., Palanichamy, M.G. and Zhang, Y-P. (2006), 'Multiple Maternal Origins of Chickens: Out of the Asian Jungles', *Molecular Phylogenetics and Evolution*, 38(1): 12–19.

Livarda, A. (2011), 'Spicing up Life in Northwestern Europe: Exotic Food Plant Imports in the Roman and Medieval World', *Vegetation History and Archaeobotany*, 20(2): 143–64.

Lobban, R.A. (1994), 'Pigs and their prohibition', *International Journal of Middle East Studies*, 26(1): 57–75.

Locker, A. (2007), '*In Piscibus Diversis*; The Bone Evidence for Fish Consumption in Roman Britain', *Britannia*, 38: 141–80.

Lokuruka, M.N.I. (2006), 'Meat is the Meal and Status is by Meat: Recognition of Rank, Wealth and Respect through Meat in Turkana Culture', *Food and Foodways*, 14: 201–29.

Longxi, Z. (1998), *Mighty Oppositions: From Dichotomies to Differences in the Comparative study of Chinas*, Stanford, CA: Stanford University Press.

Low, C. (2011), 'Birds and KhoeSān: Linking Spirits and Healing with Day-to-Day Life', *Africa*, 81(2): 295–313.

Loyn, H.R. (1970), *Anglo-Saxon England and the Norman Conquest*, London: Longman.

Lucas, G. and McGovern, T. (2007), 'Bloody Slaughter: Ritual Decapitation and Display at the Viking settlement of Hofstaðir, Iceland', *European Journal of Archaeology*, 10(1): 7–30.

Lucy, S. (2000), *The Anglo-Saxon Way of Death: Burial Rites in Early England*, Gloucester, UK: Sutton.

Ludwig, A., Pruvost, M., Reissmann, M., Benecke, N., Brockmann, G.A., Castaños, P., Cieslak, M., Lippold, S., Llorente, L., Malaspinas, A-S., Slatkin, M. and Hofreiter, M. (2009), 'Coat Color Variation at the Beginning of Horse Domestication', *Science*, 324: 485.

Luke, B. (1998), 'Violent Love: Hunting, Heterosexuality, and the Erotics of Men's Predation', *Feminist Studies*, 24(3): 627–55.

Lund, J. (2010), 'At the Water's Edge', in M. Carver, A. Sanmark and S. Semple (eds), *Signals of Belief in Early England: Anglo-Saxon Paganism Revisited*, Oxford: Oxbow.

Lyman, R.L. (1994), *Vertebrate Taphonomy*, Cambridge: Cambridge University Press.

Lyman, L. (2008), *Quantitative Palaeozoology*, Cambridge: Cambridge University Press.

Lysaght, P. (1994), *Milk and Milk Products from Medieval to Modern Times*, Edinburgh: Canongate Academic.

MacDonald, K.C. and Blend, R.M. (2000), 'Chickens', in K.F. Kiple and K.C. Ornelas (eds), *The Cambridge World History of Food Volume 1*, Cambridge: Cambridge University Press.

MacGregor, A. (1991), 'Antler, Bone and Horn', in J. Blair and N. Ramsay (eds), *English Medieval Industries: Craftsmen, Techniques, Products*, London: Hambledon Press.

MacGregor, A. (1996), 'Swan Rolls and Beak Markings: Husbandry Exploitation and Regulation of *Cygnus Olor* in England, c.1100–1900', *Anthropozoologica*, 22: 39–69.

MacKenzie, J.M. (1988), *The Empire of Nature: Hunting: Conservation and British Imperialism*, Manchester, UK: Manchester University Press.

MacKinnon, M. (2004), *Production and Consumption of Animals in Roman Italy: Integrating the Zooarchaeological and Textual Evidence*, Michigan: University of Michigan.

MacKinnon, M. (2006), 'Supplying Exotic Animals for the Roman Amphitheatre Games: New Reconstructions combining Archaeological, Ancient Textual, Historical and Ethnographic data', *Mouseion-Calgary*, 6(2): 137–61.
MacKinnon, M. (2010a), ' "Left" is "right": the symbolism behind side choice among ancient animal sacrifices', in D. Campana, P. Crabtree, S.D. deFrance, J. Lev-Tov and A.M. Choyke (eds), *Anthropological Approaches to Zooarchaeology: Colonialism, Complexity and Animal Transformations*, Oxford: Oxbow.
MacKinnon, M. (2010b), 'Cattle 'Breed' Variation and Improvement in Roman Italy: Connecting the Zooarchaeological and Ancient Textual Evidence', *World Archaeology*, 42(1): 55–73.
MacKinnon, M. (2010c), ' "Sick as a Dog": Zooarchaeological Evidence for Pet Dog Health and Welfare in the Roman World', *World Archaeology*, 42(1): 290–309.
MacKinnon, M. and Belanger, K. (2006), 'In Sickness and in Health: Care for an Arthritic Maltese Dog from the Roman cemetery of Yasmina, Carthage, Tunisia', in L.M. Snyder and E.A. Moore (eds), *Dogs and People in Social, Working Economic or Symbolic Interaction*, Oxford: Oxbow:
Madgwick, R. (2008), 'Patterns in the Modification of Animal and Human Bones in Iron Age Wessex: Revisiting the Excarnation Debate', in O. Davis, N. Sharples and K. Waddington (eds), *Changing Perspectives on the First Millennium BC: Proceedings of the Iron Age Research Student Seminar 2006*, Oxford: Oxbow.
Madgwick, R., Sykes, N., Miller, H., Symmons, R., Morris, J. and Lamb, A. (2013), 'Fallow Deer (*Dama dama dama*) Management in Roman South-East Britain', *Archaeological and Anthropological Sciences*, 5(1): 111–22.
Magennis, H. (1999), *Anglo-Saxon Appetites: Food and Drink and their Consumption in Old English and Related Literature*, Dublin: Four Courts Press.
Magennis, H. (2006), *Images of Community in Old English Poetry*, Cambridge: Cambridge University Press.
Maier, U. (1999), 'Agricultural Activities and Land Use in a Neolithic Village Around 3900 BC: Hornstaad Hörnle IA, Lake Constance, Germany', *Vegetation History and Archaeobotany*, 8(1–2): 87–94.
Mainland, I. (2008), 'The uses of Archaeological Faunal Remains in Landscape Archaeology', in B. David and J. Thomas (eds), *Handbook of Landscape Archaeology*, Walnut Creek, CA: Left Coast Press.
Maltby, M. (1990), 'Animal Bones', in J. Richards (ed.), *The Stonehenge Environs Project*, London: English Heritage.
Maltby, M. (1997), 'Domestic Fowl on Romano-British Sites; Inter-site Comparisons of Abundance', *International Journal of Osteoarchaeology*, 7: 402–14.
Maltby, M. (2006), *Integrating Zooarchaeology*, Oxford: Oxbow.
Maltby, M. (2007), 'Chop and Change: Specialist Cattle Carcass Processing in Roman Britain', in B. Croxford, N. Ray, R. Roth and N. White (eds), *TRAC 2006: Proceedings of the 16th Annual Theoretical Roman Archaeology Conference*, Oxford: Oxbow.
Mannermaa, K. (2008), 'Birds and Burials at Ajvide (Gotland, Sweden) and Zvejnieki (Latvia) about 8000–3900BP', *Journal of Anthropological Archaeology*, 27(2): 201–25.
Manning, R.B. (1993), *Hunters and Poachers: A Cultural and Social History of Unlawful Hunting in England 1485–1640*, Oxford: Clarendon Press.
Marciniak, A. (2005), *Placing Animals in the Neolithic: Social Zooarchaeology of Prehistoric Farming Communities*, London: UCL Press.
Marciniak, A. (2011), 'The Secondary Products Revolution: Empirical Evidence and its Current Zooarchaeological Critique', *Journal of World Prehistory*, 24: 117–30.

Marom, N., Bar-Oz, G. and Münger, S. (2006), 'A New Incised Scapula from Tel Kinrot', *Near Eastern Archaeology*, 69(1): 37–40.
Martin, L.A. (2000), 'Gazelle (Gazella spp.) Behavioural Ecology: Predicting Animal Behaviour for Prehistoric Environments in Southwest Asia', *Journal of Zoology*, 250: 13–30.
Marvin, G. (1984), 'The Cockfight in Andalusia, Spain: Images of the Truly Male', *Anthropological Quarterly*, 57(2): 60–70.
Marvin, G. (2000), 'The Problem of Foxes: Legitimate and Illegitimate Killing in the English Countryside', in J. Knight (ed.), *Natural Enemies: People-Wildlife Conflict in Anthropological Perspective*, London: Routledge.
Marvin, W.P. (2006), *Hunting Law and Ritual in Medieval English Literature*, Woodbridge, UK: Brewer.
Mashkour, M., Bocherens, H. and Moussa, I. (2002), 'Long Distance Movements of Sheep and Goats of Bakhtiari Nomads Tracked with Intra-tooth Variations of Stable Isotopes (^{13}C and ^{18}O)', in J. Davies, M. Fabis, I. Mainland, M. Richards and R. Thomas (eds), *Diet and Health in Past Animal Populations*, Oxford: Oxbow.
Masseti, M., Cavallaro, A., Pecchioli, E. and Vernesi, C. (2006), 'Artificial Occurrence of the Fallow Deer, *Dama dama dama* (L., 1758), on the Island of Rhodes (Greece): Insight from mtDNA Analysis', *Human Evolution*, 21(2): 167–76.
Matisoo-Smith, E. and Robins, J.H. (2004), 'Origins and Dispersals of Pacific Peoples: Evidence from mtDNA Phylogenies of Pacific Rat', *Proceedings of the National Academy of Sciences*, 101(24): 9167–72.
McCorriston, J. (1997), 'Textile Extensification, Alienation, and Social Stratification in Ancient Mesopotamia', *Current Anthropology*, 38(4): 517–35.
McCorriston, J., Harrower, M., Martin, L. and Oches, E. (2012), 'Cattle Cults of the Arabian Neolithic and Early Territorial Societies', *American Anthropologist*, 114: 45–63.
McCulloch, L. and Stancich, L. (1998), 'Women and (In)Security: The Case of the Philippines', *The Pacific Review*, 11(3): 416–43.
McCullough, M.E. and Willoughby, B.L. (2009), 'Religion, Self-regulation, and Self-control: Associations, Explanations, and Implications', *Psychological Bulletin*, 135(1): 69–93.
McGrory, S., Svensson, E.M., Götherström, A., Mulville, J., Powell, A. J., Collins, M.J. and O'Connor, T.P. (2012), 'A Novel Method for Integrated Age and Sex Determination from Archaeological Cattle Mandibles', *Journal of Archaeological Science*, 39(10): 3324–30.
Meadow, R.H. (1989), 'Osteological Evidence for the Process of Animal Domestication', in J. Clutton-Brock (ed.), *The Walking Larder: Patterns of Domestication, Pastoralism, and Predation*, London: Unwin.
Meaney, A. (1981), *Anglo-Saxon Amulets and Curing Stones* (British Archaeological Reports British Series 96), Oxford: Archeopress.
Meaney, A. (1992), 'Anglo-Saxon Idolators and Ecclesiasts from Theodore to Alcuin: A Source Study', *Anglo-Saxon Studies in Archaeology and History*, 5: 103–25.
Mennell, S. (1985), *All Manners of Food: Eating and Taste in England and France from the Middle Ages to the Present*, Oxford: Blackwell.
Mensforth, R.P. (2007), 'Human Trophy Taking in Eastern North America During the Archaic Period', in R.J. Chacon and D.H. Dye (eds), *The Taking and Displaying of Human Body Parts as Trophies by Amerindians*, New York: Springer.
Midgley, M. (1978), *Beast And Man: The Roots of Human Nature*, London: Routledge.
Miklíková, Z. and Thomas, R. (2008), *Current Research in Animal Palaeopathology: Proceedings of the Second Animal Palaeopathology Working Group Conference* (British Archaeological Reports International Series S1844), Oxford: Archaeopress.

Miletski, H. (2005), 'A History of Bestiality', in A.M. Beetz and A.L. Podberscek (eds), *Bestiality and Zoophilia: Sexual Relations with Animals*, West Lafayette, IN: Purdue University Press.
Miller, H., Cradem, R., Lamb, A., Madgwick, R., Osbourne, D. Symmons, R. and Sykes, N. (submitted), 'Dead or Alive? Investigating long-distance transport of live fallow deer and their body-parts in Antiquity', *Environmental Archaeology*.
Milliet, J. (2002), 'A Comparative Study of Women's Activities in the Domestication of Animals', in M.J. Henninger-Voss (ed.), *Animals in Human Histories: The Mirror of Nature and Culture*, Rochester, NY: University of Rochester Press.
Mills, A.C. and Slobodin, R. (1994), *Amerindian Rebirth: Reincarnation Belief Among North American Indians and Inuit*, Toronto: University of Toronto Press.
Milner, N., Craig, O.E., Bailey, G.N. and Andersen, S.H. (2006), 'A Response to Richards and Schulting', *Antiquity*, 80(308): 456–8.
Mirecki, P.A. and Meyer, M. (2002), *Magic and Ritual in the Ancient World*, Leiden: Brill.
Mithen, S.J. (1990), *Thoughtful Foragers: A Study of Prehistoric Decision Makers*, Cambridge: Cambridge University Press.
Mlekuž, D. (2007), '"Sheep are Your Mother": Rhyta and Inter-species Politics in the Neolithic of the Eastern Adriatic', *Documenta Praehistorica*, 34: 267–80.
Monbiot, G. (2013), *Feral: Rewilding the Land, the Sea, and Human Life*, Canada: Penguin
Mondini, M., Muñoz, S. and Wickler, S. (2004), *Colonisation, Migration and Marginal Areas: A Zooarchaeological Approach*, Oxford: Oxbow.
Mooney, H.A. and Hobbs, R.J. (2000), *Invasive Species in a Changing World*, Washington, DC: Island Press.
Moore, T. (2007), 'Perceiving Communities: Exchange, Landscapes and Social Networks in the Later Iron Age of Western Britain', *Oxford Journal of Archaeology*, 26(1): 79–102
Morgan, G.M. (1975), 'Three Non-Roman Blood Sports', *Classical Quarterly*, 25: 117–22.
Morphy, H. (1989), *Animals Into Art*, London: Routledge.
Morris, B. (1998), *The Power of Animals: An Ethnography*, Oxford: Berg.
Morris, B. (2000), *Animals and Ancestors: An Ethnography*, Oxford: Berg.
Morris, J.T. (2008a), *Re-examining Associated Bone Groups from Southern England and Yorkshire, c.4000BC to AD1550*. Unpublished PhD thesis, Bournemouth University.
Morris, J. (2008b), Associated Bone Groups: One Archaeologist's Rubbish is Another's Ritual Deposition', in O. Davis, N. Sharples and K. Waddington (eds), *Changing Perspectives on the First Millennium BC: Proceedings of the Iron Age Research Student Seminar 2006*. Oxford: Oxbow.
Müldner, G. and M. P. Richards (2005), 'Fast or Feast: Reconstructing Diet in Later Medieval England by Stable Isotope Analysis', *Journal of Archaeological Science*, 32: 39–48.
Müldner, G. and M.P. Richards (2006), 'Diet in Medieval England: The Evidence from Stable Isotopes', in C. Woolgar, D. Serjeantson and T. Waldron (eds), *Food in Medieval England: History and Archaeology*, Oxford: Oxford University Press.
Müldner, G.H. and Richards, M.P. (2007), 'Stable Isotope Evidence for 1500 Years of Human Diet at the City of York, UK', *American Journal of Physical Anthropology*, 133(1): 682–97.
Mullin, M.H. (1999), 'Mirrors and Windows: Sociocultural Studies of Human-Animal Relationships', *Annual Review of Anthropology*, 28: 201–24.
Mylona, D. (2008), *Fish-eating in Greece from the Fifth Century BC to the Seventh Century AD* (British Archaeological Reports, International Series 1754), Oxford: Archaeopress.
Mysterud, A. (2010), 'Still Walking on the Wild Side? Management Actions as Steps Towards 'Semi-Domestication' of Hunted Ungulates', *Journal of Applied Ecology*, 47: 920–5.

Myres, J.N.L. and Green, B. (1973), *The Anglo-Saxon Cemeteries of Caistor-by-Norwich and Markshall, Norfolk*, London: Society of Antiquaries of London.

Nadasdy, P. (2007), 'The Gift in the Animal: The Ontology of Hunting and Human–Animal Sociality', *American Ethnologist*, 34(1): 25–43.

Nanji, R and Dhalla, H. (2008), 'The Landing of the Zoroastrians at Sanjan: The Archaeological Evidence', in J. Hinnells and A. Williams (eds), *Parsis in India and the Diaspora*, Oxford: Routledge.

Nehlich, O., Fuller, B.T., Jay, M., Mora, A., Nicholson, R.A., Smith, C.I. and Richards, M.P. (2011), 'Application of Sulphur Isotope Ratios to Examine Weaning Patterns and Freshwater Fish Consumption in Roman Oxfordshire, UK', *Geochimica et Cosmochimica Acta*, 75(17): 4963–77.

Neville, J. (1999), *Representation of the Natural World in Old English Poetry*, Cambridge: Cambridge University Press.

Nicholson, A. (2011), *The Gentry: Stories of the English*, London: Harper.

Noe-Nygaard, N. (1974), 'Mesolithic Hunting in Demark Illustrated by Bone Injuries Caused by Human Weapons', *Journal of Archaeological Science*, 1: 217–48.

O'Connor, T.P. (1982), *The Animal Bones for Flaxengate, Lincoln c. 870–1500*, London: Council for British Archaeology.

O'Connor, T.P. (1997), 'Working at Relationships: Another Look at Animal Domestication', *Antiquity*, 71(271): 149–56.

O'Connor, T.P. (2000), *The Archaeology of Animal Bones*, Stroud, UK: Sutton.

O'Connor, T.P. (2001), 'On the Interpretation of Animal Bone Assemblages from *Wics*', in D. Hill and R. Cowie (eds), *Wics: The Early Medieval Trading Centres of Northern Europe*, Sheffield: Sheffield Academic Press.

O'Connor, T.P. (2007), 'Thinking About Beastly Bodies', in A. Pluskowski, (ed.), *Breaking and Shaping Beastly Bodies: Animals as Material Culture in the Middle Ages*, Oxford: Oxbow.

O'Connor, T. (2013a), 'Humans and Animals: Refuting Aquinas', *Archaeological Review from Cambridge* 28(2): 188–94.

O'Connor, T. (2013b), *Animals as Neighbours: The Past and Present of Commensal Animals*, Michigan: Michigan State University Press.

O'Connor, T. and Sykes, N. (2010), *Extinctions and Invasions: A Social History of British Fauna*. Oxford: Windgather Press.

O'Flaherty, W.D. (1980), *Women, Androgynes, and Other Mythical Beasts*, Chicago: University of Chicago Press.

Oggins, R.S. (2004), *The Kings and Their Hawks: Falconry in Medieval England*, London: Yale.

Okumura, M. and Eggers, S. (2010), 'Living and Eating in Coastal Southern Brazil during Prehistory: A Review', in D. Collard, J. Morris and E. Perego (eds), *Food and Drink in Archaeology 3*, Totnes, UK: Prospect Books.

Olsan, L. (1999), 'The Inscription of Charms in Anglo-Saxon Manuscripts', *Oral Tradition*, 14(2), 401–19.

O'Regan, H. (2002), 'From Bear Pit to Zoo', *British Archaeology*, 68: 12–19.

O'Regan, H.J. and Kitchener, A.C. (2005), 'The Effects of Captivity on the Morphology of Captive, Domesticated and Feral Mammals', *Mammal Review*, 35(3–4): 215–30.

O'Regan, H., Turner, A. and Sabin, R. (2006), 'Medieval Big Cat Remains from the Royal Menagerie at the Tower of London', *International Journal of Osteoarchaeology*, 16(5): 385–94.

Orton, D. (2010a), 'A New Tool for Zooarchaeological Analysis: ArcGIS Skeletal Templates for some Common Mammalian Species', *Internet Archaeology*, 28 (URL: http://dx.doi.org/10.11141/ia.28.4).

Orton, D. (2010b), 'Both Subject and Object: Domestication, Inalienability, and Sentient Property in Prehistory', *World Archaeology*, 42(2): 188–200.
O'Shea, J. (1989), 'The Role of Wild Resources in Small-scale Agricultural Systems: Tales from the Lakes and the Plains', in J. O'Shea and P. Halstead (eds), *Bad Year Economics*, Cambridge: Cambridge University Press.
Outram, A., Stear, N.A., Bendrey, R., Olsen, S., Kasparov, A., Zaibert, V., Thorpe, N. and Evershed, R.P. (2009), 'The Earliest Horse Harnessing and Milking', *Science*, 323(5919): 1332–5.
Overton, N. and Hamilakis, Y. (2013), 'A Manifesto for a Social Zooarchaeology: Swans and other Beings in the Mesolithic', *Archaeological Dialogues*, 20(2): 111–36
Owen-Crocker, G.R. (1986), *Dress in Anglo-Saxon England*, Manchester, UK: Manchester University Press.
Owen-Crocker, G.R, Wetherell, C. and Smith, R. (2004), *Dress in Anglo-Saxon England*, Woodbridge, UK: Boydell.
Padgett, J.M. (2004), *The Centaur's Smile: The Human Animal in Early Greek Art*, New Haven, CT: Yale University Press.
Page, R.I. (1995), *Runes and Runic Inscriptions: Collected Essays on Anglo-Saxon and Viking Runes*, Woodbridge, UK: Boydell.
Paisley, J.W and Lauer, B.A. (1988), 'Severe Facial Injuries to Infants due to Unprovoked Attacks by Pet Ferrets', *The Journal of the American Medical Association*, 259(13): 2005–6.
Pales, L. and Lambert C. (1971), *Atlas Ostéologique pour Servir à l'Identification des Mammifères du Quaternaire, I. Les Membres Herbivores*, Paris: Editions du Centre national de la recherche scientifique.
Papathanasiou, E. (2005), 'Health Status of the Neolithic Populations of Alepotrypa Cave, Greece', *American Journal of Physical Anthropology*, 126: 377–90.
Payne, A. (1990), *Medieval Beasts*, London: British Library.
Payne, S. (1973), 'Kill-off Patterns in Sheep and Goats: The Mandibles from Aşvan Kale', *Anatolia Studies*, 23: 281–303.
Pearson, J.A., Buitenhuis, H., Hedges, R.E.M., Martin, L., Russell, N. and Twiss, K.C. (2007), 'New Light on Early Caprine Herding Strategies from Isotope Analysis: A Case Study from Neolithic Anatolia', *Journal of Archaeological Science*, 34: 2170–9.
Peoples, J. and Bailey, G. (2012), *Humanity: An Introduction to Cultural Anthropology*, Belmont, CA: Wadsworth.
Perrin, O.T. (2011), 'Marks: A Distinct Subcategory within Writing as Integrationally Defined', *Language Sciences*, 33(4): 623–33.
Philo, C. and Wilbert, C. (2000), *Animal Spaces, Beastly Places: New Geographies of Human-Animal Relations*, London: Routledge
Pinault Sørensen, M. (2007), 'Portraits of Animals 1600–1800', in M. Senior (ed.) *A Cultural History if Animals in the Age of Enlightenment*, Oxford: Berg.
Pitts, M. (2007), 'The Emperor's New Clothes? The Utility of Identity in Roman Archaeology', *Journal of American Archaeology*, 111: 693–713.
Pluskowski, A.G. (2005), 'Narwhals or Unicorns? Exotic Animals as Material Culture in Medieval Europe', *European Journal of Archaeology*, 7(3): 291–313.
Pluskowski, A. (2006), *Wolves and the Wilderness in the Middle Ages*, Woodbridge, UK: Boydell.
Pluskowski, A. (2007), 'Communicating through Skin and Bones: Appropriating Animal Bodies in Medieval Western European Seigneurial Culture' in A. Pluskowski (ed.), *Breaking and Shaping Beastly Bodies: Animals as Material Culture in the Middle Ages*, Oxford: Oxbow.

Pluskowski, A. (2010), 'Animal Magic', in M. Carver, A. Sanmark and S. Semple (eds), *Signals of Belief in Early England: Anglo-Saxon Paganism Revisited*, Oxford: Oxbow.
Pluskowski, A. (2012), *The Ritual Killing and Burial of Animals: European Perspectives*, Oxford: Oxbow.
Pluskowski, A., Brown, A., Shillito, L.M., Seetah, K., Makowiecki, D., Jarzebowski, M. and Kreem, J. (2011), 'The Ecology of Crusading project: New Research on Medieval Baltic Landscapes', *Antiquity*, 85(328).
Podberscek, A.L., Paul, E.S. and Serpell, J.A. (2000), *Companion Animals and Us: Exploring the Relationship Between People and Pets*, Cambridge: Cambridge University Press.
Politis, G. and Saunders, N. (2002), 'Archaeological Correlates of Ideological Activity: Food Taboos and Spirit-animals in an Amazonian Hunter-gatherer Society', in P. Miracle and N. Milner (eds), *Consuming Patterns and Patterns of Consumption*, Cambridge: MacDonald Institute.
Pollard, J. (2006), 'A Community of Beings: Animals and People in the Neolithic of Southern Britain', in D. Serjeantson and D. Field (eds), *Animals in the Neolithic of Britain and Europe*, Oxford: Oxbow.
Ponce, P. (2010), *A Comparative Study of Activity Related Skeletal Change in 3rd-2nd Millennium BC Coastal Fishers and 1st Millennium AD Inland Agriculturalists in Chile, South America*, Unpublished PhD Thesis, Durham University, UK.
Poole, K. (2010a), 'Bird Introductions', in T. O'Connor and N. Sykes (eds), *Extinctions and Invasions: A Social History of British Fauna*, Oxford: Windgather.
Poole, K. (2010b), *The Nature of Society in England AD 410–1066*, Unpublished PhD thesis, University of Nottingham.
Poole, K. (2010c), 'Mammal and Bird Remains', in G. Thomas (ed.), *The Later Anglo-Saxon Settlement at Bishopstone: A Downland Manor in the Making*, York: Council for British Archaeology.
Poole, K. (2013), 'Horses for Courses? Religious Change and Dietary Shifts in Anglo-Saxon England', *Oxford Journal of Archaeology*, 32(3): 319–33.
Poole, K. and Lacey, E. (forthcoming), 'Avian Aurality in Anglo-Saxon England: A Case Study from Bishopstone', *World Archaeology*.
Popkin, P. (2005), 'Caprine Butchery and Bone Modification Templates: A Step Towards Standardisation', *Internet Archaeology*, 17.
Popkin, P. (2010), 'The Recognition and Interpretation of a Singular Late Bronze Age Animal Sacrifice at Kilise Tepe, Turkey', in D. Collard, J. Morris and E. Perego (eds), *Food and Drink in Archaeology 3*. Totnes, UK: Prospect Books.
Popkin, P.R.W., Baker, P., Worley, F., Payne, S. and Hammon, A. (2012), 'The Sheep Project (1): Determining Skeletal Growth, Timing of Epiphyseal Fusion and Morphometric Variation in Unimproved Shetland Sheep of Known Age, Sex, Castration Status and Nutrition', *Journal of Archaeological Science*, 39: 1775–92.
Preece, R. and Fraser, D. (2000), 'The Status of Animals in Biblical and Christian Thought: A Study in Colliding Values', *Society and Animals*, 8(3): 245–64.
Price, T.D., Knipper, C., Grupe, G. and Smrcka, V. (2004), 'Strontium Isotopes and Prehistoric Human Migration: The Bell Beaker Period in Central Europe', *European Journal of Archaeology*, 7(1): 9–40.
Prummel, W. and Niekus, M.J.L. (2011), 'Late Mesolithic Hunting of Small Female Aurochs in the Valley of the River Tjonger (the Netherlands) in the light of Mesolithic Aurochs Hunting in NW Europe', *Journal of Archaeological Science*, 38: 1456–67.
Pruvost, M., Bellone, R., Benecke, N., Sandoval-Castellanos, E., Cieslak, M., Kuznetsova, T. and Ludwig, A. (2011), 'Genotypes of Predomestic Horses Match Phenotypes Painted

in Paleolithic Works of Cave Art', *Proceedings of the National Academy of Sciences*, 108(46): 18626–30.
Puputti, A.K. (2008), 'A Zooarchaeology of Modernizing Human–Animal Relationships in Tornio, Northern Finland, 1620–1800', *Post-Medieval Archaeology*, 42(2): 304–16.
Purcell, N. (1987), 'Town in Country and Country in Town', in E.B. MacDougall (ed.), *Ancient Roman Villa Gardens*, Washington, DC: Dumbarton Oaks.
Quine, W. (1951), *It Tastes Like Chicken*, New Haven, CT: Furioso Press.
Quinn, M.S. (1993), 'Corpulent Cattle and Milk Machines: Nature, Art and the Ideal Type', *Society and Animals*, 1(2): 145–57.
Rackham, O. (2000), *The History of the Countryside: The Classic History of Britain's Landscape, Flora and Fauna*, London: Phoenix Press.
Raju, T.N. (2006), 'Continued Barriers for Breast-feeding in Public and the Workplace', *The Journal of Pediatrics*, 148(5): 677–9.
Ray, K. and Thomas, J. (2003), 'In the Kinship of Cows: The Social Centrality of Cattle in the Earlier Neolithic of Southern Britain', in M. Parker Pearson (ed.), *Food, Culture and Identity in the Neolithic and Early Bronze Age* (British Archaeological Reports 1117), Oxford: Archaeopress.
Redding, R.W. (1991), 'The Role of the Pig in the Subsistence System of Ancient Egypt: A Parable on the Potential of Faunal Data', in P.J. Crabtree and K. Ryan (ed.), *Animal Use and Cultural Change* [MASCA Research Papers in Science and Archaeology Supplement to Volume 8. MASCA. The University Museum of Archaeology and Anthropology], Philadelphia: University of Pennsylvania.
Redfern, R. (2006), *A Gendered Analysis of Health for the Iron Age to the End of the Romano-British Period in Dorset, England (Mid to Late 8th Century BC to the End of the 4th Century AD)*, Unpublished PhD thesis, University of Birmingham, UK.
Redfern, R.C. (2008), 'New Evidence for Iron Age Secondary Burial Practice and Bone Modification from Gussage All Saints and Maiden Castle (Dorset, England)', *Oxford Journal of Archaeology*, 27(3): 281–301.
Redfern, R.C. and DeWitte, S.N. (2011), 'A New Approach to the Study of Romanization in Britain: A Regional Perspective of Cultural Change in Late Iron Age and Roman Dorset using the Siler and Gompertz–Makeham Models of Mortality', *American Journal of Physical Anthropology*, 144(2): 269–85.
Reese, D. (2002), 'On the Incised Cattle Scapulae from the East Mediterranean and Near East', *Bonner Zoologische Beitrage*, 50(3), 183–98.
Rekdal, O.B. (1999), 'Cross-cultural Healing in East African Ethnography', *Medical Anthropology Quarterly*, 13(4), 458–82.
Reinken, G. (1997), 'Wieder-verbreinung. Verwendung und Namesgebungung des Damhirsches *Cervus dama* L. in Europa', *Z Jagdwiss*, 43: 197–206.
Reitz, E.J. and Wing E.S. (2008), *Zooarchaeology* (2nd edn), Cambridge: Cambridge University Press.
Restall, M. (2003), *Seven Myths of the Spanish Conquest*, Oxford: Oxford University Press.
Reynolds, A. (2009), *Anglo-Saxon Deviant Burial Customs*, Oxford: Oxford University Press.
Reynolds, H. (1981), *The Other Side of the Frontier: Aboriginal Resistance to the European Invasion of Australia*, Sydney: University of New South Wales Press.
Reynolds, R. (2013), *Food for the Soul: The Social Dynamics of Marine Fish Consumption Along the Southern North Sea Coast from AD 700 to AD 1200*, Unpublished PhD thesis, University of Nottingham, UK.

Richards, M.P. and Hedges, R.E.M. (1999), 'Stable Isotope Evidence for Similarities in the Types of Marine Foods Used by Late Mesolithic Humans at Sites Along the Atlantic coast of Europe', *Journal of Archaeological Science*, 26, 717–22.
Richards, M.P., Pearson, J.A., Molleson, T.I., Russell, N. and Martin, L. (2003), 'Stable Isotope Evidence of Diet at Neolithic Çatalhöyük, Turkey', *Journal of Archaeological Science*, 30(1): 67–76.
Richards, M.P. and Schulting, R. (2006), 'Touch Not the fish: the Mesolithic-Neolithic Change of Diet and its Significance', *Antiquity*, 80(308): 444–56.
Richards, M.P., Schulting, R.J. and Hedges, R.E. (2003), 'Archaeology: Sharp Shift in Diet at Onset of Neolithic', *Nature*, 425(6956): 366.
Richardson, A. (2012), ' "Riding like Alexander, Hunting like Diana": Gendered Aspects of the Medieval Hunt and its Landscape Settings in England and France', *Gender & History*, 24(2): 253–70.
Riches, D. (2000), 'The Holistic Person; or, the Ideology of Egalitarianism', *Journal of the Royal Anthropological Institute*, 6(4): 669–85.
Richmond, J. (2006), 'Textile Production in Prehistoric Anatolia: A Study of Three Early Bronze Age Sites', *Ancient Near Eastern Studies*, 43: 203–38.
Richter, J. (1982), 'Adult and Juvenile Aurochs, *Bos primigenius* from the Maglemosian site of Ulkestrup Lyng Øst, Denmark', *Journal of Archaeological Science*, 9(3): 247–59.
Rippon, S. (2012), *Making Sense of an Historic Landscape*, Oxford: Oxford University Press.
Ritvo H. (1987), *The Animal Estate. The English and Other Creatures in the Victorian Age*, Cambridge, MA: Harvard University Press.
Robb, J. (1997), 'Violence and Gender in Early Italy', in D.L. Martin and D.F Freyer (eds), *Troubled Times: Violence and Warfare in the Past*, New York: Gordon and Breach.
Roberts, B.K. (1987), *The Making of the English Village: A Study in Historical Geography*, Harlow, UK: Longman.
Rogers, A. (2007), 'Beyond the Economic in the Roman Fenland; Reconsidering Land, Water and Religion', in A. Flemming and R. Hingley (eds), *Prehistoric and Roman Landscapes*, Macclesfield, UK: Windgather Press.
Rosvold, J., Halley, D.J., Hufthammer, A.K., Minagawa, M. and Andersen, R. (2010), 'The Rise and Fall of Wild boar in a Northern Environment: Evidence from Stable Isotopes and Subfossil Finds', *The Holocene*, 20(7): 1113–21.
Rotherham I.D. and Lambert R.A (2011), *Invasive and Introduced Plants and Animals: Human Perceptions, Attitudes and Approaches to Management*, London: Earthscan.
Rowland, B. (1978), *Birds with Human Souls: A Guide to Bird Symbolism*, Knoxville: University of Tennessee Press.
Rowley-Conwy, P. (2000), 'Milking Caprines, Hunting Pigs', in P. Rowley-Conwy (ed.), *Animal Bones, Human Societies*, Oxford: Oxbow Press, pp. 124–32.
Rowley-Conwy, P., Albarella, U. and Dobney, K. (2012), 'Distinguishing Wild Boar from Domestic Pigs in Prehistory: A Review of Approaches and Recent Results', *Journal of World Prehistory*, 25(1), 1–44.
Ruby, M.B. and Heine, S.J. (2012), 'Too Close to Home: Factors Predicting Meat Avoidance', *Appetite*, 59(1): 47–52.
Russell, N. (2002), 'The Wild Side of Animal Domestication', *Society and Animals*, 10(3): 285–302.
Russell, N. (2007), 'The Domestication of Anthropology', in R. Cassidy and M.H. Mullin (eds), *Where the Wild Things Are Now: Domestication Reconsidered*, Oxford: Berg.
Russell, N. (2012), *Social Zooarchaeology: Humans and Animals in Prehistory*, Cambridge: Cambridge University Press.

Russell, N. and Düring, B.S. (2006), 'Worthy is the Lamb: A Double Burial at Neolithic Çatalhöyük (Turkey)', *Paléorient*, 32(1): 73–84.
Russell, T. (2013), 'Through the Skin: Exploring Pastoralist Marks and their Meanings to Understand parts of East African Rock Art', *Journal of Social Archaeology*, 13(1), 3–30.
Ryder, M.L. (1969), 'Changes in the Fleece of Sheep following Domestication', in P.J. Ucko and G.W. Dimbleby (eds), *The Domestication and Exploitation of Plants and Animals*, Chicago: Aldine Publishing.
Ryder, M.L. (1974), 'Wools from Antiquity', *Textile History*, 5: 100–10
Ryder, M.L. (1983), *Sheep and Man*, London: Duckworth.
Ryder, M.L. (2005), 'The Human Development of Different Fleece-Types in Sheep and Its Association with the Development of Textile Crafts', in F. Pritchard and J-P. Wild (eds), *Northern Archaeological Textiles NESAT VII*, Oxford: Oxbow.
Ryder, M.L. and Gabra-Sanders, T. (1985), 'The Application of Microscopy to Textile History', *Textile History*, 16(2): 123–40.
Ryder, R.D. (2000), *Animal Revolution: Changing Attitudes Towards Speciesism*, Oxford: Berg.
Sadler, P. (1990), 'The Faunal Remains', in J.R. Fairbrother (ed.), *Faccombe Netherton: Excavations of a Saxon and Medieval Manorial Complex II*. (British Museum Occasional Paper 74), London: British Museum.
Salisbury, J.E. (1994), *The Beast Within: Animals in the Middle Ages*, London: Routledge.
Salmi, A-K. (2012), 'Man's Best Friends? The Treatment of the Remains of Dogs, Cats and Horses in Early Modern Northern Finland', in Ä.T. Ikäs and A-K. Salmi (eds), *Archaeology of Social Relations: Ten Case Studies by Finnish Archaeologists*, Oulu, Finland: University of Oulu.
Salomonsson, A. (1994), 'Milk and Folk Belief: With Examples from Sweden', in P. Lysaght (ed.), *Milk and Milk Products from Medieval to Modern Times*, Edinburgh: Canongate Academic.
Sánchez Romero, M., Aranda Jiménez, G., and Alarcón García, E. (2008), 'Gender and Age Identities in Rituals of Commensality: The Argaric Societies', *Treballs d'Arqueologia*, 13: 69–89.
Saroglou, V. (2011), 'Believing, Bonding, Behaving, and Belonging: The Big Four Religious Dimensions and Cultural Variation', *Journal of Cross-Cultural Psychology*, 42(8): 1320–40.
Scandura, M., Iacolina, L. and Apollonio, M. (2011), 'Genetic Diversity in the European Wild Boar *Sus scrofa*: Phylogeography, Population Structure and Wild x Domestic Hybridization', *Mammal Review*, 41(2): 125–37.
Schlaepfer, M.A., Hoover, C. and Dodd Jr, C.K. (2005), 'Challenges in Evaluating the Impact of the Trade in Amphibians and Reptiles on Wild Populations', *BioScience*, 55(3): 256–64.
Schmid, E. (1965), Damhirsche im Römischen Augst. *Ur-Schweiz*, 29: 53–63.
Schmid, E. (1972), *Atlas of Animal Bones*, Amsterdam: Elsevier.
Schneider, J. (1987), 'The Anthropology of Cloth', *Annual Review of Anthropology*, 16: 409–48.
Schoenberger, G. (1951), 'A Goblet of Unicorn Horn', *The Metropolitan Museum of Art Bulletin*, 9(10): 284–8.
Schoeninger, M.J., DeNiro, M.J. and Tauber, H. (1983), 'Stable Nitrogen Isotope Ratios of Bone Collagen Reflect Marine and Terrestrial Components of Prehistoric Human Diet', *Science*, 220: 1381–3.

Schulting, R. (2013), 'On the Northwestern Fringes: Early Neolithic Subsistence in Britain and Ireland as seen through Faunal Remains and Stable Isotopes', in S. Colledge, S. Connoly, K. Dobney, K. Manning and S. Shennan (eds), *The Origins and Spread of Domestic Animals in Southwest Asia and Europe*, Walnut Creek, CA: Left Coast Press.
Schumacher, E.F. (1973), *Small Is Beautiful: Economics as if People Mattered*, London: Blond & Briggs.
Schuman, A. (1981), 'The rhetoric of portions', *Western Folklore*, 41(1): 72–80.
Schwabe, C.W. (1994), 'Animals in the Ancient World', in A. Manning and J. Serpell (eds), *Animals and Human Society: Changing Perspectives*, London: Routledge.
Schwarcz, H.P., Dupras, T.L. and Fairgrieve, S.I. (1999), '^{15}N Enrichment in the Sahara: In Search of a Global Relationship', *Journal of Archaeological Science*, 26: 629–36.
Scott, G.R. (1957), *The History of Cockfighting*, London: Charles Skilton.
Scully, T. (1995), *The Art of Cooking in the Middle Ages*, Woodbridge, UK: Boydell.
Searle, J.B., Kotlík, P., Rambau, R.V., Marková, S., Herman, J.S. and McDevitt, A. (2009), 'The Celtic Fringe of Britain: Insights from Small Mammal Phylogeography', *Proceedings of the Royal Society B: Biological Sciences*, 276(1677): 4287–94.
Seetah, K. (2007), 'The Middle Ages on the Block: Animals, Guilds and Meat in the Medieval Period', in A. Pluskowski (ed.), *Breaking and Shaping Beastly Bodies: Animals as Material Culture in the Middle Ages*, Oxford: Oxbow.
Seetah. K. (2008), 'Modern Analogy, Cultural Theory and Experimental Archaeology: A Merging Point at the Cutting Edge of Archaeology', *World Archaeology*, 40(1): 135–50.
Seligmann, L.J. (1987), 'The Chicken in Andean History and Myth: The Quechua Concept of Wallpa', *Ethnohistory*, 34(2): 139–70.
Semple, S. (2010), 'In the Open Air', in M. Carver, A. Sanmark and S. Semple (eds), *Signals of Belief in Early England: Anglo-Saxon Paganism Revisited*, Oxford: Oxbow.
Serjeantson, D. (2000), 'Good to Eat and Good to Think With: Classifying Animals from Complex Sites', in P. Rowley-Conwy (ed.), *Animal Bones, Human Societies*, Oxford: Oxbow.
Serjeantson, D. (2005), '"Science is Measurement"; ABMAP, A Database of Domestic Animal Bone Measurements', *Environmental Archaeology*, 10(1): 97–103.
Serjeantson, D. (2006), 'Birds as food and markers of status', in C. Woolgar, D. Serjeantson and T. Waldron (eds), *Food in Medieval England: History and Archaeology*, Oxford: Oxford University Press.
Serjeantson, D. (2007), 'Intensification of Animal Husbandry in the Late Bronze Age? The Contribution of Sheep and Pigs', in C. Hazelgrove and R. Pope (eds), *The Earlier Iron Age in Britain and the Near Continent*, Oxford: Oxbow.
Serjeantson, D. (2009), *Birds*, Cambridge: Cambridge University Press.
Serjeantson, D. (2011), *Review of Animal Remains from the Neolithic and Early Bronze Age of Southern Britain* (Research Department Report Series, 29–2011), Portsmouth, UK: English Heritage.
Serjeantson, D. and Morris, J. (2011), 'Ravens and Crows in Iron Age and Roman Britain', *Oxford Journal of Archaeology*, 30(1): 85–107.
Serjeanston, D., Wales, S. and Evans, J. (1984), 'Fish in Later Prehistoric Britain', in *Archaeo-Ichthyological Studies. Papers presented at the 6th Meeting of the ICAZ Fish Remains Working Group*, Neumünster: Wachholz Verlag.
Serjeantson, D. and Woolgar C.M. (2006), 'Fish Consumption in Medieval England', in C.M. Woolgar, D. Serjeantson, and T. Waldron (eds), *Food in Medieval England: History and Archaeology*, Oxford: Oxford University Press.
Serpell, J. (1996), *In the Company of Animals: A Study of Human–Animal Relationships*, Cambridge: Cambridge University Press.

Serpell J and Paul E. (1994), 'Pets and the Development of Positive Attitudes to Animals', in J. Serpell and A. Manning (eds), *Animals and Human Society: Changing Perspectives*, London: Routledge.
Sestieri, A.M.B. (1992), *The Iron Age Community of Osteria dell'Osa: A Study of Socio-political Development in Central Tyrrhenian Italy*, Cambridge: Cambridge University Press.
Shaw, P. (2011), *Pagan Goddesses in the Early Germanic World: Eostre, Hreda and the Cult of Matrons*, London: Bristol Classical Press.
Shear, I.M. (2002), 'Mycenaean Centaurs at Ugarit', *Journal of Hellenic Studies*, 122: 147–53.
Shelton, J-A. (2007), 'Beastly Spectacles in the Ancient Mediterranean World', in L. Kalof (ed.), *A Cultural History of Animals in Antiquity*, Oxford: Berg.
Sherratt, A. (1981), *Plough and Pastoralism: Aspects of the Secondary Products Revolution*, Cambridge: Cambridge University Press.
Sherratt, A. (1983), 'The Secondary Exploitation of Animals in the Old World', *World archaeology*, 15(1): 90–104.
Sherratt, A. (1997), *Economy and Society in Prehistoric Europe: Changing Perspectives*, Edinburgh: Edinburgh University Press.
Shettima, A.G. and Tar, U. (2008), 'Farmer-pastoralist Conflict in West Africa: Exploring the Causes and Consequences', *Information, Society and Justice*, 1(2): 163–84.
Shipman, P. (1981), *Life History of a Fossil: An Introduction to Taphonomy and Paleoecology*, Cambridge: Harvard University Press.
Shipman, P. (2010), 'The Animal Connection and Human Evolution', *Current Anthropology*, 51(4): 519–38.
Simoons, J.J. (1994), *Eat Not This Flesh: Food Avoidances from Prehistory to the Present*, London: University of Wisconsin Press.
Simoons, F.J. and Baldwin, J.A. (1982), 'Breast-feeding of Animals by Women: Its Socio-cultural Context and Geographic Occurrence', *Anthropos*, 77(3): 421–48.
Skjaervø, P.O. (2008), 'The Horse in Indo-Iranian Mythology', *Journal of the American Oriental Society*, 128(2): 295–302.
Skjelbred, A.H.B. (1994), 'Milk and Milk Products in a Women's World', in P. Lysaght (ed.), *Milk and Milk Products from Medieval to Modern Times*, Edinburgh: Canongate Academic.
Smith, M. (2006), 'Bones Chewed by Canids as Evidence for Human Excarnation: A British Case Study', *Antiquity*, 80(309): 671–85.
Sommerseth, I. (2011), 'Archaeology and the Debate on the Transition from Reindeer Hunting to Pastoralism', *Rangifer*, 31(1): 111–27.
Somvabshi, R. (2006), 'Veterinary Medicine and Animal Keeping in Ancient India', *Asian Agri-History*, 10(2): 133–46.
Speth, J.D. (1983), *Bison Kills and Bone Counts: Decision Making by Ancient Hunters*, Chicago: University of Chicago Press.
Sponheimer, M., Robinson, T., Ayliffe, L., Roeder, B., Hammer, J., Passey, B. and Ehleringer, J. (2003), 'Nitrogen Isotopes in Mammalian Herbivores: Hair δ^{15}N Values from a Controlled Feeding Study', *International Journal of Osteoarchaeology*, 13(1–2): 80–7.
Squire, D. (2012), *Chicken Nuggets: A Miscellany of Poultry Pickings*, Dartington, UK: Green Books.
Stafford, P. (1980), 'The "Farm of One Night" and the organization of King Edward's estates in Domesday', *Economic History Review*, 33: 491–502.
Stamatis, C., Suchentrunk, F., Moutou, K. A., Giacometti, M., Haerer, G., Djan, M., Vapa, L., Vukovic, M., Tvrtković, N., Sert, H., Alves, P. C. and Mamuris, Z. (2009),

'Phylogeography of the Brown Hare (*Lepus europaeus*) in Europe: A Legacy of South-eastern Mediterranean Refugia?', *Journal of Biogeography*, 36: 515–28.
Stammler, F. (2012), 'Earmarks, Furmarks and the Community: Multiple Reindeer Property among West Siberian Pastoralists', in A.M. Khazanov and G. Schlee (eds), *Who Owns the Stock?: Collective and Multiple Forms of Property in Animals*, Oxford: Berg.
Starr, R.J. (1992), 'Silvia's Deer (Vergil, Aeneid 7.479–502): Game Parks and Roman Law', *The American Journal of Philology*, 113(3): 435–9.
Steel, L. (2004), 'A Goodly Feast . . . A Cup of Mellow Wine: Feasting in Bronze Age Cyprus', *Hesperia*, 73: 281–300.
Stevens, R.E, Lister, A.M., Hedges R.E.M. (2006), 'Predicting Diet, Trophic Level and Palaeoecology from Bone Stable Isotope Analysis: A Comparative Study of Five Red Deer Populations', *Oecologia*, 149(1): 12–21.
Stevens, R.E., Lightfoot, E., Hamilton, J. Cunliffe, B. and Hedges, E.M. (2010), 'Stable Isotope Investigations of the Danebury Hillfort Pit Burials', *Oxford Journal of Archaeology*, 29(4): 407–28.
Stevens, R.E., Lightfoot, E., Allen, T. and Hedges, R.E.M. (2012), 'Palaeodiet at Eton College Rowing Course, Buckinghamshire: Isotopic Changes in Human Diet in the Neolithic, Bronze Age, Iron Age and Roman Periods through the British Isles', *Archaeological and Anthropological Science*, 4: 167–84.
Stevens, R.E., Lightfoot, E., Hamilton, J., Cunliffe, B.W. and Hedges, R.E.M. (2013), 'One for the Master and One for the Dame: Stable Isotope Investigations of Iron Age Animal Husbandry in the Danebury Environs', *Archaeological and Antrhopological Sciences*, 1: 9–109.
Stocker, D. and Stocker, M. (1996), 'Sacred Profanity: The Theology of Rabbit Breeding and the Symbolic Landscape of the Warren', *World Archaeology*, 28(2): 265–72.
Stone, D. (2006), 'The Consumption and Supply of Birds in Late Medieval England', in C. Woolgar, D. Serjeantson and T. Waldron (eds), *Food in Medieval England: History and Archaeology*, Oxford: Oxford University Press.
Storey A., Ramırez, J.M., Quiroz, D., Burley, D.V., Addison, D. J., Walter, R., Anderson, A.J., Hunt, T.L., Athens, J.S., Huynen, L. and Matisoo-Smith, E.A. (2007), 'Radiocarbon and DNA Evidence for a Pre-Columbian Introduction of Polynesian Chickens to Chile', *Proceedings of the National Academy of Sciences*, 104(25): 10335–9.
Sundkvist, A. (2004), 'Herding Horses: A Model of Prehistoric Horsemanship in Scandinavia–and Elsewhere?', in B. Santillo Frizell (ed.), *Pecus: Man and Animal in Antiquity*, Rome: the Swedish Institute of Rome.
Sutton, M.Q. (1995), 'Archaeological Aspects of Insect Use', *Journal of Archaeological Method and Theory*, 2(3): 253–98.
Swabe, J. (1999), *Animals, Disease and Human Society: Human–Animal Relations and the Rise of Veterinary Medicine*, London: Routledge.
Sykes, N.J. (2004), 'The Dynamics of Status Symbols: Wildfowl Exploitation in England AD 410–1550', *Archaeological Journal*, 161: 82–105.
Sykes, N.J. (2005), 'Hunting for the Normans: Zooarchaeology Evidence for Medieval identity', in A. Pluskowski (ed.), *Just Skin and Bones?: New Perspectives on Human–Animal Relations in the Historical Past* (British Archaeological Reports, International Series 1410), Oxford: Archaeopress.
Sykes, N.J. (2006a), 'The Impact of the Normans on Hunting Practices in England', in C. Woolgar, D. Serjeantson and T. Waldron (eds), *Food in Medieval England: History and Archaeology*, Oxford: Oxford University Press.

Sykes, N.J. (2006b), 'From Cu and Sceap to Beffe and Motton', in C. Woolgar, D. Serjeantson and T. Waldron (eds), *Food in Medieval England: History and Archaeology*, Oxford: Oxford University Press.

Sykes, N.J. (2007a), 'Taking Sides: The Social Life of Venison in Medieval England', in A Pluskowski (ed.), *Breaking and Shaping Beastly Bodies: Animals as Material Culture in the Middle Ages*, Oxford: Oxbow.

Sykes, N.J. (2007b), *The Norman Conquest: A Zooarchaeological Perspective* (British Archaeological Report, International Series 1656), Oxford: Archaeopress.

Sykes, N.J. (2007c), 'Animal Bones and Animal Parks', in R. Liddiard (ed.), *The Medieval Deer Park: New Perspectives*, Macclesfield, UK: Windgather Press.

Sykes, N.J. (2010a), 'Worldviews in Transition: The Impact of Exotic Plants and Animals on Iron Age/Romano-British landscapes', *Landscapes*, 10(2): 19–36.

Sykes, N.J. (2010b), 'Deer, Land, Knives and Halls: Social Change in Early Medieval England', *Antiquaries Journal*, 90: 175–93.

Sykes, N.J. (2010c), 'The Fallow Deer' in T. O'Connor and N. Sykes (eds), *Extinctions and Invasions: A Social History of British Fauna*, Oxford: Windgather Press.

Sykes, N.J. (2011), 'Woods and the Wild', in Hamerow, H., Hinton, D. A. and Crawford, S. (eds), *The Oxford Handbook of Anglo-Saxon Archaeology*, Oxford: Oxford University Press.

Sykes, N.J. (2012), 'A Social Perspective on the Introduction of Exotic Animals: The Case of the British Chicken', *World Archaeology*, 44(1): 158–69.

Sykes, N.J. (in press), 'Hunting and Hunting Landscapes', in C. Smith (ed.) *Encyclopaedia of Global Archaeology*, Berlin: Springer.

Sykes, N.J. and Carden, R.F. (2011), 'Were Fallow Deer Spotted (OE* pohha/* pocca) in Anglo-Saxon England? Reviewing the Evidence for *Dama dama dama* in Early Medieval Europe', *Medieval Archaeology*, 55(1): 139–62.

Sykes, N.J. and Putman, R. (in press), 'Management of Ungulates in the 21st Century: How Far Have We Come?', in R. Putman and M. Apollonio (eds), *Behaviour and Management of European Ungulates*, Dunbeath: Whittles Publishing.

Sykes, N.J. and Symmons, R. (2007), 'Sexing Cattle Horn-cores: Problems and Progress', *International Journal of Osteoarchaeology*, 17: 514–23.

Sykes, N.J., White, J., Hayes, T. and Palmer, M. (2006), 'Tracking Animals Using Strontium Isotopes in Teeth: The Role of Fallow Deer (*Dama dama*) in Roman Britain', *Antiquity*, 80: 1–12.

Sykes, N.J., Baker, K.H., Carden, R.F., Higham, T.F., Hoelzel, A.R., and Stevens, R.E. (2011), 'New Evidence for the Establishment and Management of the European Fallow Deer (Dama dama dama) in Roman Britain', *Journal of Archaeological Science*, 38(1): 156–65.

Symons, M. (2002), 'Cutting up Cultures', *Journal of Historical Sociology*, 15(4): 431–50.

Szabo, V.E. (2005), 'Bad to the Bone? The Unnatural History of Monstrous Medieval Whales', *Heroic Age: A Journal of Early Medieval Northwestern Europe*, 8: 1–18.

Szynkiewicz, S. (1990), 'Sheep Bone as a Sign of Human Descent: Tibial Symbolism among the Mongols', in R. Willis (ed.), *Signifying Animals: Human Meaning in the Natural World*, London: Routledge.

Tani, Y. (1996), 'Domestic Animal as Serf: Ideologies of Nature in the Mediterranean and Middle East', in R. Ellen and K. Fukui (eds), *Redefining Nature: Ecology, Culture and Domestication*. Oxford: Berg.

Tapper, R. (1998), 'Animality, Humanity, Morality, Society', 47–62 in Ingold T., *What is an Animal?*, Unwin: Routledge.

Tapper, R.L. (1994), 'Animality, Humanity, Morality, Society', in T. Ingold (ed.), *What is an Animal?*, London: Routledge.

Taylor G. (2000), *Castration: An Abbreviated History of Western Manhood*, London: Routledge.

Teegen, W.-R. (2005), 'Rib and Vertebral Fractures in Medieval Dogs from Haithabu, Starigard and Schleswig', in J. Davies, M. Fabis, I. Mainland, M. Richards and R. Thomas (eds), *Diet and Health in Past Animal Populations*, Oxford: Oxbow.

Thomas, J. (2004), 'Current Debates on the Mesolithic-Neolithic Transition in Britain and Ireland', *Documenta Praehistorica*, 31, 113–30.

Thomas, J. (2012), 'Archaeologies of Place and Landscape', in I. Hodder (ed.), *Archaeological Theory Today* (2nd edn), Cambridge: Polity Press.

Thomas, K. (1983), *Man and the Natural World: Changing Attitudes in England 1500–1800*, London: Penguin.

Thomas, R. (2005a), 'Zooarchaeology, Improvement and the British Agricultural Revolution', *International Journal of Historical Archaeology*, 9(2): 71–88.

Thomas R. (2005b), 'Perception Versus Reality: Changing Attitudes Towards Pets in Medieval and Post-medieval England', in A. Pluskowski (ed.), *Just Skin and Bones? New Perspectives on Human-Animal Relationships in the Historical Past* (British Archaeological Rerports International Series 1410), Oxford: Archaeopress.

Thomas, R. (2005c), *Animals, Economy and Status: Integrating Zooarchaeological and Historical Data in the Study of Dudley Castle, West Midlands (c. 1100–1750)* (British Archaeological Report, British Series 392), Oxford: Archaeopress.

Thomas R. (2007), 'Maintaining Social Boundaries through the Consumption of Food in Medieval England', in Twiss K (ed.), *The Archaeology of Food and Identity* [Occasional Publication No. 34], Carbondale, IL: Center for Archaeological Investigations.

Thomas, R. (2009), 'Bones of Contention: Why Later Post-medieval Assemblages of Animal Bones Matter', in A. Horning and M. Palmer (eds), *Crossing Paths or Sharing Tracks: Future Directions in the Archaeological Study of Post-1550 Britain and Ireland*, Woodbridge, UK: Boydell.

Thomas, R. (2010), 'Translocated *Testudinidae*: The Earliest Archaeological Evidence for Tortoises in Britain', *Post-Medieval Archaeology*, 44(1): 165–71.

Thomas, R. (2012), 'Nonhuman Paleopathology', in J. Buikstra and C. Roberts (eds), *The Global History of Paleopathology: Pioneers and Prospects*, Oxford: Oxford University Press.

Thomas, R., and McFadyen, L. (2010), 'Animals and Cotswold-Severn Long Barrows: A Re-examination', *Proceedings of the Prehistoric Society*, 76: 95–113.

Thomas, R., Holmes, M. and Morris, J. (2013), '"So Bigge as Bigge May Be": Tracking Size and Shape Change in Domestic Livestock in London (AD 1220–1900)', *Journal of Archaeological Science*, 40(8): 3309–25.

Thorn, F. and Thorn, C. (1986), *Domesday Book, 25, Shropshire*, Chichester, UK: Phillimore.

Tieszen, L.L. and Fagre, T. (1993), 'Effect of Diet Quality and Composition on the Isotopic Composition of Respiratory CO^2, Bone Collagen, Bioapatite, and Soft Tissues', in G. Grupe and J.B. Lambert (eds), *Prehistoric Human Bone*, Berlin: Springer-Verlag.

Tixier-Boichard, M., Bed'hom, B. and Rognon, X. (2011), 'Chicken Domestication: La Domestication du Poulet: de l'Archéologie à la Génomique', *Comptes Rendus Biologies*, 334(3): 197–204.

Tolan-Smith, C. (2008), 'Mesolithic Britain', in G. Bailey and P. Spikkins (eds), *Mesolithic Europe*, Cambridge: Cambridge University Press.

Towers, J., Montgomery, J., Evans, J., Jay, M. and Parker Pearson, M. (2010), 'An Investigation of the Origins of Cattle and Aurochs Deposited in the Early Bronze Age Barrows at Gayhurst and Irthlingborough', *Journal of Archaeological Science*, 37(3): 508–15.

Toynbee, J.M. (1971), *Death and Burial in the Roman World*, London: Cornell University Press.
Tracey, J. (2012), 'New Evidence for Iron Age Burial and Propitiation Practices in Southern Britain', *Oxford Journal of Archaeology*, 31(4): 367–79.
Treherne, P. (1995), 'The Warrior's Beauty: The Masculine Body and Self-identity in Bronze-Age Europe', *Journal of European Archaeology*, 3(1): 105–44.
Trut, L.N. (1999), 'Early Canid Domestication: The Farm-Fox Experiment', *American Scientist*, 87(2): 160–69.
Tsoukala, V. (2009), 'Honorary Shares of Sacrificial Meat in Attic Vase Painting: Visual Signs of Distinction and Civic Identity', *Hesperia*, 78(1): 1–40.
Tsurushima, H. (2007), 'The Eleventh Century in England through Fish-eyes: Salmon, Herring, Oysters, and 1066', *Anglo-Norman Studies*, 29: 193–213.
Twiss, K. (2012), 'The Archaeology of Food and Social Diversity', *Journal of Archaeological Research*, 20(4): 357–95.
Upex, B. and Dobney, K. (2011), 'More Than Just Mad Cows: Exploring Human-Animal Relationships Through Animal Paleopathology', in A.L. Grauer (ed.), *A Companion to Paleopathology*, Chichester, UK: Wiley-Blackwell.
Ünal, A. (1988), '"You should build for eternity" New Light on the Hittite Architects and Their Work', *Journal of Cuneiform Studies*, 40: 97–106.
Urban, H.B. (2011), *The Church of Scientology: A History of a New Religion*. Princeton, NJ: Princeton University Press.
Van der Veen, M. (2003), 'When Is Food a Luxury?', *World Archaeology*, 34(3): 405–27.
Van der Veen, M. (2008), 'Food as Embodied Material Culture: Diversity and Change in Plant Food Consumption in Roman Britain', *Journal of Roman Archaeoogy*, 21: 83–110.
Van der Veen, M., Livarda, A. and Hill, A. (2008), 'New Plant Foods in Roman Britain: Dispersal and Social Access', *Environmental Archaeology*, 13(1): 11–35.
Van Groenigen, J.W. and Van Kessel, C. (2002), 'Salinity-induced Patterns of Natural Abundance Carbon-13 and Nitrogen-15 in Plant and Soil,' *Soil Science Society of America Journal*, 66(2): 489–98.
Vanhaeren, M. and d'Errico, F. (2005), 'Grave Goods from the Saint-Germain-la-Rivière Burial: Evidence for Social Inequality in the Upper Palaeolithic', *Journal of Anthropological Archaeology*, 24(2): 117–34.
Van Itterbeeck, J. and van Huis, A. (2012), 'Environmental Manipulation for Edible Insect Procurement: A Historical Perspective', *Journal of Ethnobiology and Ethnomedicine*, 8(3): 1–7.
Vann, S. (2008), 'Animal Palaeopathology at two Roman sites in Southern Britain', in Z. Miklíková and R. Thomas (eds), *Current Research in Animal Palaeopathology: Proceedings of the Second Animal Palaeopathology Working Group Conference* (British Archaeological Reports International Series S1844), Oxford: Archaeopress.
Vann, S. and Thomas, R. (2006), 'Humans, Other Animals and Disease: A Comparative Approach Towards the Development of a Standardised Recording Protocol for Animal Palaeopathology', *Internet Archaeology*, 20.
Veenhof, K.R. (1972), *Aspects of Old Assyrian Trade and Its Terminology*, Leiden: Brill.
Veraina, L. (2013), 'Excarnation and the City: The Tower of Silence Debates in Mumbai', in I. Becci, M. Burchardt and J. Casanova (eds), *Topographies of Faith*, Leiden: Brill.
Vigne, J-D. (2011), 'The Origins of Animal Domestication and Husbandry: A Major Change in the History of Humanity and the Biosphere', *Comptes Rendus Biologies*, 334(3): 171–81.

Vigne, J-D. and Helmer, D. (2007), 'Was Milk a "Secondary Product" in the Old World Neolithisation Process? Its Role in the Domestication of Cattle, Sheep and Goats', *Anthropozoologica*, 42(2): 9–40.

Viner, S., Evans, J., Albarella, U. and Parker Pearson, M. (2010), 'Cattle Mobility in Prehistoric Britain: Strontium Isotope Analysis of Cattle Teeth from Durrington Walls (Wiltshire, Britain)', *Journal of Archaeological Science*, 37(11): 2812–20.

Vitebsky, P. (2006), *The Reindeer People: Living with Animals and Spirits in Siberia*, London: Harper Collins.

Vogel, J.C.,and Van der Merwe, N.J. (1977), 'Isotopic Evidence for Early Maize Cultivation in New York State', *American Antiquity*, 42: 238–42.

von den Driesch, A. (1976), *A Guide to the Measurement of Animal Bones from Archaeological Sites*, Cambridge. MA: Peabody Museum.

Walker, R. (1985), *A Guide to Post-cranial bones of East African Animals: Mrs Walker's Bone Book*, Norwich, UK: Hylochoerus.

Walker-Meikle, K. (2012), *Medieval Pets*, Woodbridge, UK: Boydell.

Wallis, F. (1999), *Bede: The Reckoning of Time*, Liverpool, UK: Liverpool University Press.

Wapnish, P. (1995), 'Towards Establishing a Conceptual Basis for Animal Categories in Archaeology', in D.B. Small (ed.), *Methods in the Mediterranean: Historical and Archaeological Views on Texts and Archaeology*, Leiden: Brill.

Wapnish, P. and Hesse, B. (1993), 'Pampered Pooches or Plain Pariahs? The Ashkelon Dog Burials', *The Biblical Archaeologist*, 56: 55–80.

Watkins, C. (1995), *How to Kill a Dragon: Aspects of Indo-European Poetics*, Oxford: Oxford University Press.

Wells, P.S. (2011), 'The Iron Age', in S. Milisauskas (ed.), *European Prehistory: A Survey*, London: Springer.

Wessing, R. (1988), 'Spirits of the Earth and Spirits of the Water: Chthonic Forces in the Mountains of West Java', *Asian Folklore Studies*, 47(1): 43–6.

Wessing, R. (1995), 'The Last Tiger in East Java: Symbolic Continuity in Ecological Change', *Asian Folklore Studies*, 4(2): 191–218.

Wessing, R. (2006), 'Symbolic Animals in the Land Between the Waters: Markers of Place and Transition', *Asian Folklore Studies*, 6: 205–39.

Wessing, R. and Jordaan, R.E. (1997), 'Death at the Building Site: Construction Sacrifice in Southeast Asia,' *History of Religions*, 37(2): 101–21.

West, B. and Zhou, B-X. (1988), 'Did Chickens go North? New Evidence for Domestication', *Journal of Archaeological Science*, 15: 515–33.

West, M.L. (2007), *Indo-European Poetry and Myth*, Oxford: Oxford University Press.

White, C.D., Pohl, M.E., Schwarcz, H.P. and Longstaffe, F.J. (2001), 'Isotopic Evidence for Maya Patterns of Deer and Dog use at Preclassic Colha', *Journal of Archaeological Science*, 28(1): 89–107.

Whitehouse, R. (2012), 'Gender in Central Mediterranean Prehistory', in M.L.S. Sørensen (ed.), *The History of Gender Archaeology in Northern Europe: A Companion to Gender Prehistory*, Cambridge: Polity Press.

Whittle, A.W. (1996), *Europe in the Neolithic: The Creation of New Worlds*, Cambridge: Cambridge University Press.

Whittle, A. (2007), 'Going Over: People and Their Times', in A. Whittle and V. Cummings (eds), *Going Over: The Mesolithis-Neolithic Transition in North West Europe*, Oxford: Oxford University Press.

Whittle, A. (2012), 'Dead or alive? Grand narratives and intimate histories of (mainly Neolithic) animals', URL: http://www.alexandriaarchive.org/bonecommons/items/show/1646.

WHO/IUCN/WWF (1993), *Guidelines on Conservation of Medicinal Plants*, Gland, Switzerland: IUCN.
Wickham-Jones, C.R. (2010), *Fear of Farming*, Oxford: Windgather Press.
Wiedemann, F., Bocherens, H., Mariotti, A., von den Driesch, A. and Grupe, G. (1996), 'Methodological and Archaeological Implications of Intra-tooth Isotopic Variations (d[13]C, d[18]O) in Herbivores from Ain Ghazal (Jordan, Neolithic)', *Journal of Archaeological Science*, 26: 697–704.
Wild, J.P. (2003a), 'Anatolia and the Levant in the Neolithic and Chalcolithic period, *c.* 8000–3500/3300 BC', in D.T. Jenkins (ed.), *The Cambridge History of Western Textiles 1*, Cambridge: Cambridge University Press.
Wild, J.P. (2003b) 'Romans in the West, 600 BC – AD 400', in D.T. Jenkins (ed.), *The Cambridge History of Western Textiles 1*, Cambridge: Cambridge University Press.
Wilkie, R.M. (2010), *Livestock/Deadstock: Working with Farm Animals from Birth to Slaughter*, Philadelphia: Temple University Press.
Willerslev, R. (2007), *Soul Hunters: Hunting, Animism, and Personhood Among the Siberian Yukaghirs*. Berkeley: University of California Press.
Williams, H. (2007), 'An Ideology of Transformation: Cremation Rites and Animal Sacrifice in Early Anglo-Saxon England', in N. Price (ed.), *The Archaeology of Shamanism*, London: Routledge.
Willis, S. (2007), 'Sea, Coast, Estuary, Land and Culture in Iron Age Britain', in C. Haselgrove and T. Moore (eds), *The Later Iron Age in Britain and Beyond*, Oxford: Oxbow.
Wilson, C.A. (2003), *Food and Drink in Britain from the Stone Age to the 19th Century*, Chicago: Academy Chicago.
Winter, W. (1997), 'A Lone Loanword and its Implications', in S. Eliasson and E.H. Jahr (eds), *Language and its Ecology: Essays in Memory of Einar Haugen*, Berlin: Walter de Gruyter.
Wolch, J. and Emel, J. (1998), *Animal Geographies: Place, Politics and Identity in the Nature-Culture Borderlands*. London: Verso.
Woodward, P. and Woodward, A. (2004), 'Dedicating the Town: Urban Foundation Deposits in Roman Britain', *World Archaeology*, 36(1): 68–86.
Wright J.C. (2004), *The Mycenaean Feast*, Oxford: Oxbow.
Yalden, D. (1999), *A History of British Mammals*, London: T. & A. Poyser
Yalden, D. (2010), 'Conclusion', in T. O'Connor and N. Sykes (eds), *Extinctions and Invasions: A Social History of British Fauna*, Oxford: Windgather Press.
Yannouli, E. and Trantalidou, K. (1999), 'The Fallow Deer (*Dama dama* Linnaeus, 1758): Archaeological Presence and Representation in Greece', in N. Benecke (ed.), *The Holocene History of the European Vertebrate Fauna*, Rahden/Westf: Verlag Marie Leidorf GmbH.
Yapp, B. (1981), *Birds in Medieval Manuscripts*, London: The British Library.
Yates, D.T. (2007), *Land, Power and Prestige. Bronze Age Field Systems in Southern England*. Oxford: Oxbow.
Yentsch, A. (1991), 'Engendering Visible and Invisible Ceramic Artifacts, Especially Dairy Vessels', *Historical Archaeology*, 25(4): 132–55.
Yvinec, J.-H. (1993), 'La Part du Gibier dans l'Alimentation du Haut Moyen Âge', in J. Desse and F. Audoin-Rouzeau (eds), *Rencontres Internationales d'Archéologie et d'Histoire d'Antibes (13e: 1992: Ville d'Antibes): Exploitation des Animaux Sauvages à Travers le Temps*, Juan-les-Pins: APDCA.
Zander, A. (2013), *An Examination of the Faunal and Lithic Assemblages of the European Mesolithic Sites with Antler Frontlets*, Unpublished BA dissertation, University of Nottingham, UK.

Zeder, M. (1991), *Feeding Cities: Specialized Animal Economy in the Ancient Near East*, Washington, DC: Smithsonian Institute Press.

Zeder, M.A. (2005), 'A View From the Zagros: New Perspectives on Livestock Domestication in the Fertile Crescent', in J-D. Vigne, J. Peters and D. Helmer (eds), *The First Steps of Animal Domestication: New Archaeozoological Approaches*, Oxford: Oxbow.

Zeder, M.A. (2006), 'Reconciling Rates of Long Bone Fusion and Tooth Eruption and Wear in Sheep (*Ovis*) and Goat (*Capra*)', in D. Ruscillo (ed.), *Ageing and Sexing Animals from Archaeological Sites*, Oxford: Oxbow.

Zeder, M.A. (2011a), 'Pathways to Animal Domestication', in A. Damania and P. Gepts (eds), *Harlan II: Biodiversity in Agriculture: Domestication, Evolution, & Sustainability*, Davis: University of California.

Zeder, M.A. (2011b), 'The Origins of Agriculture in the Near East', *Current Anthropology*, 53(4): 221–35.

Zeuner, F. (1963), *A History of Domesticated Animals*, London: Hutchinson.

Zhang, H., Paijmans, J.L., Chang, F., Wu, X., Chen, G., Lei, C. and Hofreiter, M. (2013), 'Morphological and Genetic Evidence for Early Holocene Cattle Management in Northeastern China', *Nature Communications*, 4. doi:10.1038/ncomms3755.

Zohary, D., Tchernov, E. and Horwitz, L. (1998), 'The Role of Unconscious Selection in the Domestication of Sheep and Goats', *Journal of Zoology*, 245(2): 129–35.

Zvelebil, M. (2008), 'Innovating Hunter-gatherers: The Mesolithic in the Baltic', in G. Bailey and P. Spikkins (eds), *Mesolithic Europe*, Cambridge: Cambridge University Press.

INDEX